THE APPLICATIONS OF THE SCIENCE OF REINCARNATION

THE APPLICATIONS OF THE SCIENCE OF REINCARNATION

A Scientific Explanation of the Afterlife, Aliens, Disclosure, and Our Current Reality That Even Homer Simpson Could Understand

Book 3
The Matrix of Consciousness Series

By Bob Good

Boynton Beach Fl

© 2025 by Bob Good

IASOR Press

Boynton Beach, Florida 33436

All rights reserved. This book or any portion thereof may not be reproduced or used in any manner whatsoever without the express written permission of the publisher except for the use of brief quotations in a book review.

ISBN: TASOR ISBN Numbers
ISBN: 978-1-7351185-6-7 for print
ISBN: 978-1-7351185-7-4 for electronic media

To send correspondence: info@iasor.org

To my wife, who gave me a life

**The Matrix of Consciousness Series
Book 3
The Applications of the Science of Reincarnation**

The Matrix of Consciousness Series
Book 1 The Science of Reincarnation
Book 2 The Mathematics of the Science of Reincarnation
Book 3 The Applications of the Science of Reincarnation
Book 4 The Science of Reincarnation College Level Textbooks

Other Books by Bob Good
The Reincarnation Series—Science Fiction
Currently Dead
The Reincarnation Strategy
The Lies We Tell Ourselves

Contents

Introduction xi

PART ONE 1
    The Problem
CHAPTER 1 3
    Extraterrestrial Aliens
CHAPTER 2 13
    Transdimensional Aliens
CHAPTER 3 19
    The Gateway Process: The Science of the Soul
CHAPTER 4 51
    The Afterlife
CHAPTER 5 83
    Humanity's Current Situation: It Is More
    Complex Than You Think
CHAPTER 6 107
    Not Disclosure but Context
CHAPTER 7 131
    Disclosure: It Has Happened

PART TWO 169
    The Solution
CHAPTER 8 171
    The Institute for Advanced Consciousness
    Studies

CHAPTER 9 209
    The Consciousness Proposal
CHAPTER 10 243
    Education
CHAPTER 11 267
    Applied Research
CHAPTER 12 285
    Theory Development
CHAPTER 13 321
    Empiricism
CHAPTER 14 329
    Army Futures Command: Military Analysis
        of the Data and Resultant Strategy
CHAPTER 15 351
    TIFACS Policy Positions
CHAPTER 16 379
    The Common Constitution
CHAPTER 17 401
    Army Futures Command: Religion
        and Exopolitics
CHAPTER 18 417
    Conclusion

About the Author 453

The Applications of the Science of Reincarnation

Introduction

How should a reader approach a book titled *The Applications of the Science of Reincarnation?* The expectation would be to receive information, evidence supporting that information, and guidance on how to apply it.

As an author, my job is to make simple a very complex topic and tell a story. This is not a dry science book; it is a story that explains what is happening in consciousness science. It contains the afterlife, aliens, CIA cover-ups, and the clarity of truth. I want it simple and entertaining enough so anyone can understand it, and the very brightest among us can deal with the complexity of the problems that confront us all.

I am not asking you to suspend disbelief but rather look at the information I present before making your judgment. This is not so much about supplying proof as it is about expanding your understanding to the limits of what science has learned and what the government is suppressing.

After graduating college in 1971, I became a helicopter pilot in the US Army, and at certain points during that journey, I was a test pilot with a top-secret security clearance. After the service and for the next forty years, I was an independent contractor in the pharmaceutical industry, specializing in lab animal research. Also, during that time, I was, for a while, in an organization called Business Executives for National Security. Their mission statement was to bring business solutions to military problems.

That is what this book is about—bringing business solutions to military, political, and social problems.

The problem, in one sentence, is that there is now an intersection between heaven and aliens. The afterlife is a dimensional wall that we pierce when we shed our bodies and before coming back with new ones—the same dimensional wall transdimensional aliens go through on their journeys on and around Earth.

In this new emerging paradigm, we encounter two categories of aliens: extraterrestrials and transdimensionals. Can you see the government's many problems? We are going to explain these problems and present logical global solutions.

Interestingly, the US government acknowledges all this if you read the right declassified documents and know which pictures to ask for from the Jet Propulsion Laboratory.

Whether you choose to believe that the government has proven the existence of heaven or not is not the point. The point I am going to prove is that they believe it and are acting on and dealing with that reality. They just do not know what to do about it. What is worse is that the aliens are influencing them, you, and us.

We can contact alien species who will help us, but we must do this together.

This book is divided into two parts: first, outlining the new conditions that are emerging, and then proposing the best possible solution for humanity. I name names and dollar amounts and, to the best of my ability, explain the science behind the discoveries. If this is of interest to you, come and join me on the journey. Let us begin with extraterrestrials.

Part One
The Problem

Chapter 1

Extraterrestrial Aliens

People tend to forget events and do not connect them, leading to a lack of understanding. So, we will begin this narrative in 1962, just as America is entering the Space Age. I recommend watching *Secret Space UFOs: In the Beginning*, a documentary on Prime, as a beginning for those who want to see the people involved and what they said. It makes the following points.

Beginning with the X-15 program, NASA began detecting alien vehicles. On July 17, 1962, Major Robert White flew the X-15 to an altitude of 314,750 feet, or 59 miles, becoming the first "winged astronaut." He was the first to fly at Mach 4, Mach 5, and Mach 6, as well as the first to pilot a winged vehicle into space.

This was the first X-15 flight that qualified the pilot for a United States Air Force (USAF) astronaut rating. During the flight, he made the following transmission: "There are things out there! There absolutely is."[1] (Time

---

[1] *Secret Space UFOs: In the Beginning*, directed by Darcy Weir, release date May 1, 2022, Occult Journeys, 1 hr. 25 min., https://www.imdb.com/title/tt14998318/.

stamp 15:41 in the movie, and you can watch the video of the UFO.)

In the August 3, 1962, issue of *Life* magazine, he talked about his UFO encounter[2] (time stamp 16:35). The film is from the rear-facing camera on the X-15.

This was not the only UFO encounter during the X-15 program. Several months earlier, Joseph Walker's X-15 flight captured similar images.

On May 11, 1962, at the Second National Conference on the Peaceful Uses of Space Research in Seattle, Washington, Walker also claimed that one of the mission objectives during his April 1962 flight was to detect and film UFOs at high altitudes. So, at that point, NASA was actively tracking UFOs.

In May 1963 Gordon Cooper made twenty-two flights around the Earth in a Mercury-Atlas 9 spacecraft. Over Perth, Australia, he reported a greenish object with a red tail approaching him. He maintained that the object was piloted because of the maneuvers it made toward him and then away from him.

During 1965 and 1966, NASA astronauts flew a series of two-man missions into low Earth orbit to gain the experience needed to make the upcoming Apollo missions a success. In twenty months, sixteen astronauts conducted a total of ten missions in the Gemini spacecraft. Next to

---

[2] *Secret Space UFOs: In the Beginning*, directed by Darcy Weir, release date May 1, 2022, Occult Journeys, 1 hr. 25 min., https://www.imdb.com/title/tt14998318/.

the pilot's seat, on the right side of the spacecraft, the VTR was to verbally document any visual observations made during the flight.

On June 3, 1965, during the Gemini 4 mission, James McDivitt claimed he saw something unusual. "It had big arms sticking out of it...I got pictures with the movie camera."[3]

This is at timestamp 2;03 of *Secret Space UFOs: In the Beginning*. Ground radar indicated there should not have been another object near them.

On December 4, 1965, during a Gemini 7 mission, Frank Borman reported a bogey. "We have a bogey at ten o'clock high."[4] (This is at time stamp 22:29.)

Also, in 1965 the Russians launched the Zond 3 spacecraft. This satellite took many shots of the far side of the moon, looking back at Earth.

The Zond 3 Soviet moon probe took a picture of the moon, and clearly depicted in that picture is a large tower. They referred to it as the Tower of Babel. This is not a natural structure. Frame 25 of that structure is clear enough to estimate the height of the tower to be 3.5 miles tall. Smaller structures can be seen nearby.

---

[3] *Secret Space UFOs: In the Beginning*, directed by Darcy Weir, release date May 1, 2022, Occult Journeys, 1 hr. 25 min., https://www.imdb.com/title/tt14998318/.

[4] *Secret Space UFOs: In the Beginning*, directed by Darcy Weir, release date May 1, 2022, Occult Journeys, 1 hr. 25 min., https://www.imdb.com/title/tt14998318/.

Neither the Russians nor NASA have an explanation for this object.

In early 2000 a whistleblower named Karl Wolf claimed to have worked for the Air Force as a photography technician at Langley Air Force Base. He processed things such as gun camera film.

At the time, the first Lunar Orbiter program was looking for landing sites for the 1969 lunar mission. Wolf was asked by his superior to look at a printer that was having difficulty printing at an NSA facility on base. The NSA was bringing in telemetered information from radio telescopes all over the world to process that data into pictures of the moon.

Wolf, an airman second class at the time, was working with another airman second class who was charged with the care of the troublesome printer. This was a point in time when they used printed circuit boards. In his conversations with the other airman second class, the airman said, "You know we have discovered a base on the back side of the moon."[5] (Time stamp 23:47) It was information the airman should not have shared, but we were just finding out the extent of this alien base, and many in the intelligence service would have been overwhelmed by the significance of the find.

Intelligence agencies had Ingo Swann remote-view the moon and the alien base. His descriptions of these

---

[5] *Secret Space UFOs: In the Beginning*, directed by Darcy Weir, release date May 1, 2022, Occult Journeys, 1 hr. 25 min., https://www.imdb.com/title/tt14998318/.

events can be found in his book *Penetration*. They had Ingo Swann remote-view Jupiter before the Voyager 1 and Voyager 2 missions. They had Joe McMoneagle remote-view Mars one million years ago. I will explain remote viewing later in this book, but for now, imagine targeted clairvoyance—think of something and then actually look at it.

It was our knowledge of the aliens on the moon that fueled the space race, not our desire to go there, as Kennedy said, but to chase the aliens.

We already knew they were here and had absolute proof in the form of alien vehicles and aliens themselves or their bodies thereof. The crash in Roswell on July 7, 1947, occurred because the aliens were snooping around the only nuclear site in America's arsenal at that time.

These were the small Grays. The US government recovered four bodies. I want to begin to differentiate the various groups of aliens and their agendas.

The world's first nuclear explosion occurred on July 16, 1945, when a plutonium implosion device was tested at a site located 210 miles south of Los Alamos, New Mexico, on the plains of the Alamogordo Bombing Range. Roswell is 140 miles from Alamogordo and 225 from Los Alamos.

If we follow the facts and disclosures over the last seventy-five years, the fact that aliens are here and have been here is irrefutable.

While NASA has stage-managed information given

to the public and doctored photographs, it was Ingo Swann's remote viewing excursion to the moon that opened space for every one of us. We all can remote-view.

This story is not told as proof of the alien presence, which at this point is incontrovertible, but rather their intent and scope of the alien presence. This book and plan are the countermeasures humans should take—not to confront this alien presence but rather to improve and enhance the situation for everyone's benefit: humans, extraterrestrials, and extradimensional races.

These alien visitors are here on earth now, and technology transfer has occurred and is occurring along with DNA manipulation of the human species. This is already scientifically verifiable as it is recorded in our DNA, each one of us.

The policy of secrecy within the intelligence services is now detrimental and is putting us all at risk, including our aviators. All of us who look know what the intelligence services are not saying: that aliens are here, and space is filled with life. That there are multiple alien groups, each with its own agenda, and some outcomes do not favor humanity.

Governments have lied and covered up alien contact, and there are remediating protocols to the lies like CE5 protocols, Mutual UFO Network (MUFON), and NGOs that remote-view. These narratives align with government lies to produce the most probable narrative. This plan presented here is a best response to government inaction and lies. This same government has tried to protect us. This plan is designed to help us both and incorporate the best results for humanity.

For the government agencies charged with protecting us and the divergent alien groups here competing for resources, this proposal outlines a course of action that is in all our best interests at the intersection of our needs and desires.

- To face these threats, humanity must look across its myths and precognitions and see the reality that surrounds us, and this awakening will be psychically painful as old beliefs give way to new understanding.
- The purpose of this book is to fund research into these issues to protect both the Earth and humanity. The central point all these issues revolve around is human consciousness.
- The only path through the maze is the development of human consciousness. This book strategizes options for best results. This is where we need to spend the money.
- It will take collective courage, and adversaries must become allies. We either succeed or fail together.

At certain points in a battle, leadership can become overwhelmed. We are at such a point in human development. We are seeing the universe open to us if it does not crush us first. It is for us to change to meet its demands. So, a cohesive plan should be implemented to address this threat/opportunity—in a word, crisis.

The plan:

Who pays for it?
Who designs it?
Who implements it?
What is the process for modifying it for future generations?

The following outlines that plan.

It is a plan to collect and distribute an initial fundraising effort of $450 million for consciousness research. To be clear at the very beginning, this money will fund research in areas of consciousness following a fact-based and logic-driven analysis. That fact-based analysis posits the following, supported by the hard scientific evidence presented here:

- There is a continuity of our consciousness after our permanent bodily death.
- We can be trained to exit our bodies and visit this space while we are alive.
- We are conscious in both three-dimensional (3D) space and fourth-dimensional (4D) space, through which we remote-view and psychics see.

- Both these spaces are occupied by other intelligent life-forms, and they are now, and have been, interacting with humanity for thousands of years.

I want to close this chapter not by raising an alarm but rather by urging you to calm down. Aliens have been here for at least twelve thousand years, and in some cases, their society is half a million years more developed than ours. Think of this as yourself when you were very young and found out there was no Santa Claus. If they wanted to hurt you, they would have done so already. Let us wake up and deal with this.

Chapter 2

Transdimensional Aliens

The second category of aliens is called nonbiological entities (NBE). These are intelligent energy forms that are self-directed and detectable. But what I offer is more than proof. This is not only the government's acknowledgment but also the process by which any one of us can access this dimensional permeability. Once we cross that dimensional boundary for either self-enlightenment or governmental intent, the rules change. The area is too vast to comprehend, but we can navigate this space and transcend time.

The proof that what I am telling you is true comes from the Department of the Army in a document written in 1983 and declassified by the CIA in 2003. The document is titled *Analysis and Assessment of Gateway Process* and can be found in the CIA Reading room.[6]

The first line of the conclusion on page 26 reads as follows: "There is a sound, rational basis in terms of physical

---

[6] Lt. Col. Wayne M. McDonnell, *Analysis and Assessment of The Gateway Process*, Department of the Army, 1983, https://www.cia.gov/readingroom/docs/cia-rdp96-00788r001700210016-5.pdf.

science parameters for considering Gateway to be plausible in terms of its essential objectives."

The real aha moment here occurs on page 27, the final page. The report's author, Lt. Col. Wayne M. McDonnell, lists the steps to be followed. Here are the two of them:

1. "Be intellectually prepared to react to possible encounters with intelligent, non-corporeal energy forms when time-space boundaries are exceeded." These are aliens without bodies. That is what he is saying. The US Army and the CIA are stating that the Gateway Process is real, and when using it, this is to be expected.
2. "Arrange to have groups of people in Focus 12 state [a level of consciousness within Gateway] unite their altered consciousness to build holographic patterns around sensitive areas to repulse possible unwanted out-of-body presences."

This recommendation validates the military's and intelligence communities' belief in the efficacy of the Gateway Process. The military is using this process to defend our national secrets against acknowledged NBEs.

Our dimension is not a single dimension but many, and through the Gateway Process, we can access past lives. Whatever heaven you wish, you can talk to the dead, have out-of-body experiences, meet and talk to aliens, and remotely view far-off places.

We access it with a scientifically defined second body tethered to our material body. It is made of electrical energy. I will explain this more fully in chapter 3. Any being with an EKG reading has this ability. Some would call this second body a soul; it is the same. As scientists, we do not want the religious overlay. But as we will see, souls of a specific energy—or religion, if you will—aggregate together in the afterlife. Scientifically, what you will find is that all religions are the same—none is better than the other. In terms of gaining the afterlife, even if you do not believe in any religion, you will. In that regard, religion is unimportant in gaining the afterlife.

In later chapters, we will explore this second body and give you conclusive proof that you, too, can contact our alien neighbors.

Everyone can access this new dimensional space. This space is inhabited by aliens who, like us, are out of body and directly connected to our 3D reality. We can be there and encounter aliens who lack physical bodies in that space but, like us, can connect to their bodies here. With this explanation, we just proved aliens exist in our time and space.

While this serves as validation on its own, there is so much more. You see, the exopolitics—the politics surrounding aliens in our present—may be quite troubling to some, but it is nothing to be afraid of. They have been here for tens of thousands of years and know all about us; I would argue they know more about us than we do

The Applications of the Science of Reincarnation

ourselves. This rudimentary explanation of exopolitics leads directly to earth politics.

For context, we did not know about microbes until the 1500s. They were all around us throughout our history. Some were dangerous; some were helpful. We learned about them and studied them, and we became more knowledgeable. That will happen here. There is no reason to be afraid.

Now that this has all been explained, it should change how we govern ourselves, incorporating the realization that there are aliens in our present. To be accepted as an intergalactic species by others, we must show the maturity necessary for them to welcome us as participants. A condition of that acceptance is that we are not a threat to them; we must demonstrate good intentions. This means we must stop killing each other. This may seem like a big task for such a small book, but this case must be made. It must be made to our leaders and each other. This change must be funded globally, and this entire effort can be made by funding the science of reincarnation.

But funding the science of reincarnation is not just altruistic; it is also a strategic initiative in our joint national defense. For us to evidence growth as a species and as a culture to alien races thousands of years more advanced than us, we must subjugate our beliefs to our common survival. This step will be noticed by aliens who study us daily. Additionally, this funding cannot come solely from our governments but must also include contributions from our richest citizens—billionaires. It is in their interest to do so. Without a healthy population prepared to defend this planet, the billionaires are nothing more than insignificant bugs, easily swept away by races far older than our own.

At the end of the day, this is a business proposal to the aliens, to our leaders, and to ourselves to unify an effort to save us and our planet. We propose forming a new nongovernmental organization called TIFACS: the Institute for Advanced Consciousness Studies.

TIFACS will be a global organization of scientists. Politicians will not and do not understand this new landscape. Input and understanding will come from those who navigate and access this new dimensional landscape. The intent here is to benefit everyone, as we are all in this together. There are no losers with this proposal. This will be made clear as this explanation unfolds.

While TIFACS is an NGO, it is foundational that general information and access to dimensional travel procedures be made available to the public. This, in turn, leads to contact with extradimensional intelligence, and this exopolitical contact leads to recommendations for earth politics to position humanity for intergalactic contact with alien species. These are military outcomes.

Now that we have painted a rather complex puzzle, let us prove this to my readers' satisfaction, one step at a time.

This will constitute disclosure that aliens exist by our

authorities, a smooth, seamless explanation that validates belief systems and allays any fears of threat or societal upheaval. This small book intends to accomplish that to the benefit of all parties concerned.

To do that we must address a new military strategy where enemies must become allies to face a larger external threat. We will start with scientists, followed by billionaires, the commanding general of the Army Future Corp, and finally politicians proposing a common solution to a common problem.

To quote Benjamin Franklin, "We must all hang together, or we will surely all hang separately."

So how do we address our belief in the afterlife?

How do we navigate that space since the government quite correctly has demonstrated any of us can?

How do we deal with possible alien intervention from a 4D platform in our society?

If these alien societies have been here from our inception, there is less to fear and more to understand.

How we govern ourselves would preclude our ability not only to understand this but also to respond to it in a meaningful way. In short, we have been divided and conquered.

Chapter 3

The Gateway Process: The Science of the Soul

To explain the Gateway Process, its significance, and its acceptance by the military and intelligence services, you must first understand dimensional permeability. This concept is not hard to understand. The first part involves dimensions—science has well documented the existence of other dimensions. Permeability means the ability to go through. It is porous to some stuff and will block others. It is like filter media. If you have ever used a strainer in a kitchen, you get it.

Your energy can pierce that filter media, but your body cannot. We all can do this—pierce that filter media so we can see what is in the fourth dimension. Religion says that is where your soul goes, even calling your energy a soul. Science in the Gateway Process calls that energy your second body.

So, what does it look like when you look through this veil, and what can this process be called? We see this process manifest in our lives as clairvoyance, premonitions, remote viewing, and out-of-body experiences (OBEs). They are all part of quantum biology, the intersection of electricity and matter—both of which comprise you.

When you die and flatline, that is a definition of death. Your electricity has gone to the afterlife.

To understand the afterlife scientifically, what people call your soul is pure energy—the same energy that shows up in an EKG. This energy sees this 3D world through its body. While the body cannot go through the dimensional barrier, the energy can.

Some of us have a natural ability to be clairvoyant, to have premonitions that when the phone rings, you know Nana has died before you answer it.

Robert Monroe began having OBEs at a very young age. As he matured, he sought out scientists and conducted experiments to understand this ability and define its properties. I have said so far that your soul is electromagnetic, but there is something more we now know. We have not discovered the property of our electromagnetic awareness that holds us in our bodies and transfers us to new ones. So, for now, I will refer to your soul as electromagnetic because of its behavior and hold out that X unknown property that we know for certain exists.

We do know that it is somehow related to electromagnetism. As Monroe said this property is to magnetism as magnetism is to electricity as electricity is to this currently unknown property.

I will make the case that your electromagnetic signature detected by an EKG machine is your soul. Now this is simplistic because something is operating that we do not understand when we get into the science of it. But we are beginning to understand, and like the example I gave about not having all the pieces of a jigsaw puzzle together but still seeing the picture, I will leave it to you to make your own determination after I present my case.

After we go through this, we will discuss what we need to do about this emerging model of your consciousness.

In the next chapter, I will explain the fractal math proof of life beyond death and describe the process of the period souls go through between lives. We have also shown the upload and download process to be a fractal of magnitude in measuring data transmission, meaning the size of your personal data can be and is encoded in that energy. Finally, death is described as a flat line, meaning bodily death is when the electricity leaves the body.

Where, then, is the scientific data that shows our soul is indeed ours and we study it? Most, if not all, of the study of reincarnation centers on more esoteric things such as meditation and prophets—until Robert Monroe. The following are excerpts from Wikipedia about Robert Monroe:

> Robert Allan Monroe (October 30, 1915–March 17, 1995) was an American radio broadcasting executive who became known for his ideas about altered states of consciousness and for founding The Monroe Institute which continues to promote

those ideas. His 1971 book *Journeys Out of the Body* is credited with popularizing the term "out-of-body experience."

According to his own account, while experimenting with sleep-learning in 1958 Monroe experienced an unusual phenomenon, which he described as sensations of paralysis and vibration accompanied by a bright light that appeared to be shining on him from a shallow angle. Monroe went on to say that this occurred another nine times over the next six weeks, culminating in his first out-of-body experience (OBE). Monroe recorded his account in his 1971 book *Journeys Out of the Body* and went on to become a prominent researcher in the field of human consciousness. Monroe later authored two more books on his experiments with OBE, *Far Journeys* (1985) and *Ultimate Journey* (1994).

Monroe developed Hemi-Sync which he claimed could facilitate enhanced brain performance.

In 1975, Monroe registered the first of several patents concerning audio techniques designed to stimulate brain functions until the left and right hemispheres became synchronized. Monroe held that this state, dubbed Hemi-Sync, could be used to promote mental well-being or to trigger an altered state of consciousness. Monroe's concept was based on an earlier hypothesis known as binaural beats and has since been expanded upon on a commercial basis by the self-help industry.

Hemi-Sync is short for *Hemispheric Synchronization,* also known as *brainwave synchronization.* Monroe indicated that the technique synchronizes the two hemispheres of one's brain, thereby creating a "frequency-following response" designed to evoke certain effects. Hemi-Sync has been used for many purposes, including relaxation and sleep induction, learning and memory aids, helping those with physical and mental difficulties, and reaching altered states of consciousness through the use of sound.[7]

This is how the Monroe Institute describes itself on its home page:

The world's leading education center for the study of human consciousness.

> For over 50 years the Monroe Institute has been welcoming consciousness explorers from all over the world. Our nondogmatic experiential approach allows you to pursue your own personal exploration of human consciousness. Monroe

---

[7] "Robert Monroe," Wikipedia, https://en.wikipedia.org/wiki/Robert_Monroe.

programs include the use of Monroe Sound Science along with exercises to target specific states of consciousness.

This is how the Monroe Institute is described on Wikipedia:

> The Monroe Institute (TMI) is a nonprofit education and research organization devoted to the exploration of human consciousness, based in Faber, Virginia, United States. Upwards of 20,000 people are estimated to have attended TMI's residential Gateway program during its first thirty years. TMI claims a policy of no dogma or bias with respect to belief system, religion, political or social stance. The institute is housed in several buildings on 300 acres (1.2 km^2) of land south of Charlottesville, Virginia, USA.
> In 1978, the U.S. military evaluated TMI and arranged to send officers there for OBE training. In 1983, it sent additional officers.[8]

The *Analysis and Assessment of Gateway Process*, the document I quoted in the introduction, was written in the year 1983 and declassified in 2003.

---

[8] "Robert Monroe," Wikipedia, https://en.wikipedia.org/wiki/Robert_Monroe.

The Gateway Process: The Science of the Soul

We are going to make a very complex subject as simple as we can. According to the Monroe Institute and Robert Monroe, you have a second body. It has characteristics that can be enumerated using Robert Monroe's list. The government of the United States considered this information both valid and worthy of secrecy, classifying the analysis secret until it was declassified in 2003.

"First, this Second Body has weight as we understand it. It is subject to gravitational attraction although much less than the physical body"[9]

"Second, this Second Body is visible under certain conditions. To be visible, it must either reflect or radiate light in the known spectrum or at least a harmonic in this area."

"Third, the sense of touch in the Second Body seems to be very similar to that in the physical."[10]

"Fourth, the Second Body is very plastic and may take whatever form is suitable to or desired by the individual."

"Fifth, there exists the possibility the Second Body is a direct reversal of the physical."

"Sixth, direct investigation tends to support the premise of a connecting 'cord' between the

---

[9] Robert Monroe, *Journeys Out of the Body* (Crown Publishing Group, a division of Random House LLC 1971, 1979), 176.

[10] Robert Monroe, *Journeys Out of the Body* (Crown Publishing Group, a division of Random House LLC, 1971, 1979), 177.

physical and the Second Body, as described many times through the ages in esoteric literature."[11]

Now what we are going to posit here is this: The electricity measured in your body is your soul or your second body; they are the same. When you die, that electricity running through your microtubules leaves your body, and without that energy you are dead. But you are alive in what is called Locale II.

That energy is you, not your body. It can reach outside your body when you are alive.

Stephan Schwartz and I would discuss this in the early days of TIFACS when Stephan, Dean Radin, and I were trying to figure out how to come up with $450 million for consciousness research. He tells me that the electricity in your body is not who you are; it is something else yet undiscovered. Let us call it X for a moment. Robert Monroe tells me that this force we will call X is to electricity, as electricity is to magnetism as magnetism is to X. The evidence would support Monroe's opinion.

Monroe states when describing the characteristics of the second body:

"Seventh, the relationship between the second body and electricity and electromagnetic fields is quite

---

[11] Robert Monroe, *Journeys Out of the Body* (Crown Publishing Group, a division of Random House LLC, 1971, 1979), 178.

significant. The experiment in the Faraday cage points to this."[12]

So where do you go when you leave your body? Again, Robert Monroe traveled these dimensions and laid out his experience, breaking it into three Second State environments.

"Locale I is the most believable. It consists of people and places that exist in the material and well-known world at the very moment of the experiment. It is a world represented to us by our physical senses which most of us are fairly sure does exist."[13] This is a place where Robert Monroe went when he began to have OBEs. It is the here and now, and in this place, Locale I, he could appear in Los Angeles simply by his intent, even though his physical body was in New York City. According to Monroe,

> The best introduction to Locale II is to suggest a room with a sign over the door saying, "Please Check All Physical Concepts Here."...*Postulate:* Locale II is a non-material environment with laws of motion and matter only remotely related to the physical world. It is an immensity whose bounds are unknown (to this experimenter), and has

---

[12] Robert Monroe, *Journeys Out of the Body* (Crown Publishing Group, a division of Random House LLC, 1971, 1979), 178.

[13] Robert Monroe, *Journeys Out of the Body* (Crown Publishing Group, a division of Random House LLC, 1971, 1979), 60.

depth and dimension incomprehensible to the finite, conscious mind. In this vastness lie all the aspects we attribute to heaven and hell, which are but a part of Locale II. It is inhabited, if that is the word, by entities with various degrees of intelligence with whom communication is possible...Locale II is the natural environment of the Second Body.

This means both heaven and hell exist here. Note that we are seeing fractals repeating across a wide variety of instances. The repetitive event is the visual math proof across the same narratives of near-death experiences, children who remember prior lives, past-life regression, and now this. But this is different. With the other examples, we could not go into the next dimension. Now we can, and it has to do with brain wave frequency.

Locale II consists of images you would recognize, even though it is not material and exists solely as "thought." Monroe's comments about these images include the following:

> The purpose seems to be that of simulation of the physical environment—temporarily, at least—for the benefit of those just emerging from the physical world, after "death." This is done to reduce trauma and shock for the "newcomers" by

introducing familiar shapes and settings in the early conversion stages.[14]

The stories of near-death experiencers and children who remember prior lives describe similar conditions, which will be discussed in the next chapter.

This process is the same for all religions, as death is the same for all of us, regardless of which religion we follow. In that regard, all religions are part of a common denomination. This means all religious laws should be changed to remove apostasy, blasphemy, and celibacy from all religions.

This is for the religion's protection. An example of change in action is that any religion that does not educate its women is dooming itself to extinction. Take the Taliban, from swords to IEDs and missiles. With this emerging science, if the women are not educated, they cannot help the men in this increasingly complex situation. This is something the Taliban must come to itself, but pointing it out to the younger generations will speed the transition.

According to Monroe, "Locale III, in summary, proved to be a physical-matter world almost identical to our own. The natural environment is the same. There are trees, houses, cities, people, artifacts, and all the appurtenances

---

[14] Robert Monroe, *Journeys Out of the Body* (Crown Publishing Group, a division of Random House LLC, 1971, 1979), 75.

of a reasonable civil society. However, more careful study showed that it can be neither the present nor the past of our physical-matter world."

So, the Gateway Process is what Monroe used to navigate these dimensions. Here is how it works.

Here's Lt. Col. Wayne M. McDonnell's explanation of the Gateway Process, as detailed in the *Assessment and Analysis of Gateway Process*:

> Focus levels:
> The time has come to examine the specific techniques which comprise the Gateway Training Process. These techniques are designed to enable the user of the Gateway tapes to manipulate the high energy states which can be achieved if the user continues to work with the tapes over a period of time.[15]

The Gateway Process starts here, following McDonnell's explanation. It begins with your physical preparation, like beginning a meditation session. You calm yourself and focus your attention on the process, slowing your breathing and eliminating extraneous noise or distractions. The Gateway Process starts like this, with its protocol, listening to the Gateway tapes.

---

[15] Lt. Col. Wayne M. McDonnell, *Analysis and Assessment of The Gateway Process*, Department of the Army, 1983, 19, https://www.cia.gov/readingroom/docs/cia-rdp96-00788r001700210016-5.pdf.

First, you isolate all your extraneous concerns and put them mentally in a visualization of something called an energy conversion box.

Next, you achieve a state of resonance by humming the way monks hum, low and slow, on the cadence of your breath. This is called resonant tuning. This is a single monotone of protracted humming that sets up a feeling of vibration, particularly in the head. At this point you are humming along with a chorus of such sounds as are contained on the Gateway tapes. Next, you recite affirmations to the effect that you are more than your physical body, and you deeply desire to expand your consciousness.

After that, you are exposed for the first time to the Hemi-Sync sound frequencies. This is progressive and systematic, designed to put the body on the virtual threshold of sleep. This is also designed to calm the left hemisphere of the brain while raising the attentiveness of the right hemisphere.

Once that is complete, you are asked to envisage an "energy balloon" made up of an energy flow that begins at the center of your head, resembling a fountain, spiraling down your entire body, and being reabsorbed through the bottom of your feet. You are inside this energy balloon. It is designed to provide protection against conscious entities possessing lower energy levels, which you might encounter in the event you achieve an OBE involving a direct projection outside the terrestrial sphere.

So, what is happening to you as you do this?

What you are doing is Hemi-Sync induction. The following is from Wikipedia.

> The technique involves using sound waves to entrain brain waves. Wearing headphones, Monroe claimed that brains respond by producing a third sound (called binaural beats) that encouraged various brainwave activity changes. In 2002, a University of Virginia presentation at the Society for Psychophysiological Research examined Monroe's claim. The presentation demonstrated that EEG changes did not occur when the standard electromagnetic headphones of Monroe's setup were replaced by air conduction headphones, which were connected to a remote transducer by rubber tubes. This suggests that the basis for the entrainment effects is electromagnetic rather than acoustical.[16]

The entrainment effect is electromagnetic rather than acoustical. I want to repeat that because I want to reinforce the point that the soul or second body is composed of and responds to electromagnetic influences as defined by these experiments.

---

[16] "Robert Monroe," Wikipedia, https://en.wikipedia.org/wiki/Robert_Monroe.

Electromagnetism travels at a variety of frequencies.

Now what is happening to you, the subject, physiologically?

There are five main types of brain waves: alpha, beta, delta, gamma, and theta. Each one is associated with a different state of mind and frequency of consciousness.

> There are five widely recognized brain wave states characterized by their frequency bands: gamma (> 35 Hz; when you concentrate), beta (12–35 Hz; when you have a busy active mind), alpha (8–12 Hz; when you are reflective and restful), theta (4–8 Hz; when you are deeply relaxed or dreamy), and delta (0.5–4 Hz; when you are unconscious).[17]

How to activate alpha brain waves?

Meditation or exercise.

Deep breathing and closed-eye visualization are techniques that mindfulness meditation usually employs to boost alpha waves. Besides relaxation, alpha waves may also help boost creativity. They also act as a natural antidepressant by promoting the release of the neurotransmitter serotonin.

---

[17] Response to "What are the five type of brain waves and their frequencies?" Google AI Overview. 1/15/2025

What do beta brain waves do?

Beta waves are high-frequency, low-amplitude brain waves that are commonly observed in an awakened state. They are involved in conscious thought and logical thinking. They tend to have a stimulating effect.

What do delta brain waves do?

Delta waves are the slowest recorded brain waves in human beings. They are found most often in infants and young children and are associated with the deepest levels of relaxation and restorative, healing sleep. Delta is prominently seen in brain injuries, learning problems, inability to think, and severe ADHD.

What do theta brain waves do?

They are seen in connection with creativity, intuition, daydreaming, and fantasizing and are a repository for memories, emotions, and sensations. Theta waves are strong during internal focus, meditation, prayer, and spiritual awareness.

Theta waves are present during hypnosis, deep meditation, and light sleep, including the all-important REM dream state. It is the realm of your subconsciousness.

What do gamma brain waves do?

A brain producing lots of gamma waves reflects complex neural organization and heightened awareness. Gamma is associated with very high levels of intellectual function, creativity, integration, peak states, and flow states.

Can brain waves be manipulated?

People can, for example, switch their brain state into relaxation or concentration with a light-and-sound machine; they can train their brain waves to cure their ADHD or solve their sleeping problems with a neurofeedback device.

Does listening to frequencies work?

Studies have shown that binaural beats can reduce anxiety by 26.3 percent. They also aid with sleep: Certain frequencies of binaural beats activate specific brain waves, helping the body and mind relax for deeper sleep.

So, where Monroe has taken us is the region of theta waves. It is a natural state we all go through during periods of sleep. It is something monks work for thirty years to achieve. Monroe's method is faster and more targeted.

In 1994 a front-page article in *The Wall Street Journal* reported confirmation from the former director of the Intelligence and Security Command of the US Army sending personnel to the institute. It also stated the opinion of the head of the Zen Buddhist temple in Vancouver, British Columbia, who stated, "Gateway students can reach meditation states in a week that took [me] thirty years of sitting."[18]

What exists there is a human perception that we all have—that when in this state, we can seek information

---

[18] "Robert Monroe," Wikipedia, https://en.wikipedia.org/wiki/Robert_Monroe.

that to us would be available in no other way. They are states of being or levels of focus, depending on where you look.

Focus 10: "Having reached Focus 10, the participant is now ready to endeavor to achieve a state of sufficiently expanded awareness to begin actually interacting with dimensions beyond those associated with his experience of physical reality."[19]

Focus 12: Additional forms of pink and white noise enter the sound stream, and the participant begins to deploy a series of tools, including patterning, color breathing, and the energy bar tool.

Focus 15: "Travel into the past involves further expansion of consciousness through the inclusion of additional levels of sound on the Hemi-Sync tapes. The Monroe Institute trainers affirm that with enough practice, eventually Focus 15 can be achieved." Focus 21: The Future. "The last and most advanced of all the Focus states associated with the Gateway training program involves movement outside of the boundaries of time-space as in Focus 15 but with attention to discovering the future rather than the past. The individual who has achieved this state has reached a truly advanced level."[20]

"The Out-of-Body Movement: Only one tape...which

---

[19] Lt. Col. Wayne M. McDonnell, *Analysis and Assessment of The Gateway Process*, Department of the Army, 1983, 20, https://www.cia.gov/readingroom/docs/cia-rdp96-00788r001700210016-5.pdf.

[20] Lt. Col. Wayne M. McDonnell, *Analysis and Assessment of The Gateway Process*,

makes up the Gateway Experience is devoted to…out-of-body movement. This technique employs Beta signals of around 2877.3 CPS (cycles per second). Thirty to forty CPS is to be considered the normal range for Beta brainwave signals associated with the wakeful state."[21]

Brain wave frequency is directly related to dimensional perception.

Information collection potential refers to the ability to extract correct information from an event in the past shrouded in mystery or predict a future event, a technique that would clearly be valuable to intelligence services globally.

Now we understand the information pathways that are available to us. These scientific discoveries were classified at first, known and used by very few. But over time this information leaked out—look to the assessment and analysis report, written in 1983 and declassified twenty years later, as well as the acceleration of technology and the internet.

Today, in categories on Reddit such as "r/gatewaytapes," "r/aliens," "r/ufos," "r/astralprojection," and "r/experiencers," among so many others, normal people interact with a wide variety of alien species many thousands

---

Department of the Army, 1983, 22, https://www.cia.gov/readingroom/docs/cia-rdp96-00788r001700210016-5.pdf.

[21] Lt. Col. Wayne M. McDonnell, *Analysis and Assessment of The Gateway Process*, Department of the Army, 1983, 22, https://www.cia.gov/readingroom/docs/cia-rdp96-00788r001700210016-5.pdf.

of years more advanced than our own. The following is a post from Reddit where the person posting is communicating with an individual from the alien species called the Tall Whites. The poster is doing this through the system explained above.

The First Time I Met a Tall White, and What He Taught Me
This happened last year, in November.

So, I was up early in morning, I found out I didn't have work that morning. And I decided to meditate, just visualizing meditating in my astral body in outer space. A being somehow approaches me from the backdrop of space, asking to connect, so I did, and I felt that this was a very tall, skinny, pale, kindly, and intelligent being, taller than the Greys I know. I asked what this being was curious about, and the being humbly asked me for some data.

He said; "I would like to see human interaction with their pets, if that's all right?"

And gestured to my cat near my body. I looked at my kitty, and the being said,

"Yes!"

Emphatically, as there was some communication errors between us. So I pet my cat, and conveyed to the being how I understand my cat's cues, and how we share space and affection for one another. The being was very pleased.

He said, "It is the same! We call our dear animal companions differently, but the relationship dynamic is the same as humans have for their 'pets.' This is good data. Thank you for sharing."

He was very grateful to me. And I was happy to help. I decided to ask him if I could have information from him in exchange for my help, and he agreed easily, saying, "Ask anything of me."

I first asked to see him more clearly, and he agreed readily. I teleported to his location with my astral body and sensed I was on a dimly lit ship of some kind. I was immediately struck by his appearance. He was easily double my height (I'm just over 5 ft tall) and white other than his eyes. Not like a white human being, he Glowed white. His skin shined a white color, and he had a bright, emanating white aura about two to three inches from his body. His eyes were dark, but I didn't get a good look at them. He was, however, tall and lanky akin to the Greys I normally met (6 ft–7 ft tall).

I then asked him about his group, in the context of how it relates to other groups in space, around earth at that moment. I told him that a lot of humans on earth were confused about the dynamics of different beings around earth, and that if he could give me any clarifying information to communicate with my fellow humans, it would help a lot because there is a lot of confusion there.

He agreed eagerly, and said, "Oh, I can help with that easily."

He showed me a circle and said, "Imagine that this circle represents a defining line, separating the inner part, where ETs that are interested in all Earth has to offer are, and the outer area outside the circle where ETs that have no interest in Earth at all are."

I visualized the circle, and he added that the line itself is what separated the groups, noting that groups inside the circle, but closer to the line, would be less interested in Earth than groups farther into the center of the circle.

Then he said, "Now visualize a very small circle, in the center of the large one, that is Earth."

And I did that. Then he projected the circle high above me, with the Earth circle still inside it, he made it huuuuge, like a planetarium projection/display above my head. He said, "Now I will add groups."

And the large circle representation filled with a bunch of small ovals, representing ET groups, some clustered near Earth, others clustered to the outer area of the big circle, even more were outside the circle.

I asked him how many ovals (groups) were outside the circle. He laughed and said,

"A truly endless amount, there is no end to all that are out there to be found."

So, I looked at all the ovals, and I asked him about the ones clustered around Earth, marked by a line from Earth to them, and he said,

"The ones close to Earth communicate that they walk the Earth in some sense, or have outposts on Earth, or otherwise interact with humans in some direct or physical way."

I asked about one oval that overlapped with the Earth, and he said, "Those are the 'reptilian like group' native to Earth; they overlap because Earth is also their home."(The phrase is in parentheses, because he was borrowing my terminology for them, "Saurians.")

Then he said,

"Now, I will show you categories to distinguish groups."

He created 4 distinctions, and made the whole representation a 3D transparent sphere within a sphere with 3D ovals in my mind's eye;

1. ETs interacting with humans for humanity's benefit,
2. ETs interacting with humans for their own benefit,
3. ETs interacting with or studying Earth life other than humans,
4. ETs studying or interacting with the Earth itself.

He then projected these distinctions onto the

big sphere, and the majority of the 3D ovals were sorted in these categories, from closest to Earth to farthest, with the most of them being between the first two distinctions.

He showed that a sizable amount of groups were still not even here for humans, but just for animals, plants, the earth, etc.

He also showed that some groups were in several or more than one distinction at the same time.

I asked him what group was his, where did his group fall?

And he showed me that his group existed in the distinctions between 1 and 3, halfway from Earth to the outer sphere.

I asked him how his group was that way, and he explained, that, in his civilization, it is known that all living beings occupy niches and environments that overlap with the surrounding life-forms.

He explained that in a forest, all creatures overlap areas and lifestyles around each other, and that in encountering and approaching or seeing each other, intrinsic boundaries are communicated between life-forms.

He taught me that every interaction with other creatures is actually a valuable communication that teaches other life-forms my boundaries, even if I haven't realized or haven't intentionally tried to do that.

So, he continued, that his civilization, in being aware of that, lives harmoniously with many other life-forms on their planet.

He explained that humans are actually already very close to living in harmony with other animals, since so many animals already share urban environments with humans, even if humans don't like that or don't see that.

He explained that his group's goal, his goal, is to learn enough about humans, and Earth animals, to teach and show humans how to live with, and how to maintain living with, other creatures in a safe, harmonious, and non-polluting, natural way.

And to that end, he is learning about humans and their pets.

I dwelled on the sphere graph/diagram he was still showing me, and I asked him why some, but only a small number, of 3D ovals inside the circle, didn't fall into any of the distinctions he made.

He laughed and explained that some ETs are here to study ETs, not humans, and not the Earth.

I was very surprised by this, but he laughed again and said, some beings, or civilizations, just don't prioritize humans, or studying humans at all.

I understood him, and I remarked how surprising that was.

He said I was right to be surprised though, since these groups barely ever interacted with humans, and humans would only, if ever, see these groups in the backgrounds of ships, interacting with other ETs.

I asked him what type of things these outlier groups are studying, and he replied, "Some of what (things) they are here for, are/is so abstract, you physically cannot imagine it, even if I tried to show you."

I shared my amazement with him for that, how incredible that they study or interact with things I cannot comprehend, only vaguely related to Earth in any way.

And he agreed with me, explaining that it is an inevitable situation for a civilization as young as humanity.

I conveyed my gratitude to him, and I asked him if I could know his name (if he had one) to honor him and this knowledge he shared and give other humans a future possibility of meeting him.

He showed me a concept of some kind of physical principle or constant, that keeps separation between distinctly different things, like matter, vibrations, complexity, etc.

I didn't quite understand, so he looked through my memories, to see what I knew, and he said his name, was "Surface Tension." I thanked Surface Tension for his time, and he thanked me for mine, and both he and I let the connection end.[22]

This is not a stand-alone example, and here we must go back to our intelligence services, which are resource-weak. We can fairly ask them, "How are you monitoring and organizing NGO activity regarding the continued communication between humans and aliens when there is a public lack of acknowledgment that we are in contact with alien civilizations now? How are you funding additional research into something there is official obfuscation about?"

Let us look at the double-edged sword this chapter is about and continue the information collection potential of this information through the words of Nikola Tesla:

> Alpha waves in the human brain are between six and eight hertz. The wave frequency of the human cavity resonates between six and eight hertz. All biological systems operate in the same frequency range. The human brain's alpha waves function in this range and the electrical resonance of the earth is between six and eight hertz. Thus, our entire biological system—the brain and the

---

[22] forbiddensnackie, "The First time I met a Tall White, and what he taught me," Reddit, November 4, 2021, https://www.reddit.com/r/Experiencers/comments/1cvctau/the_first_time_i_met_a_tall_white_and_what_he/?share_id=YC8RMZ4ExY9Ce8PNWMaV2&utm_content=1&utm_medium=ios_app&utm_name=ioscss&utm_source=share&utm_term=10.

earth itself—work on the same frequencies. If we can control that resonate system electronically, we can directly control the entire mental system of humankind.[23]

What Tesla described is a cloud. Not for your computer's information but yours. In this cloud are your memories, thoughts, and the very essence of your personality. In short, you.

We accept the upload and download of information for our computer, but that same fractal process we cannot accept scientifically for ourselves despite believing religiously. If it is there, then why can't we access it? In fact, we all can.

This is where the rubber meets the road. The fundamental connection between you, the afterlife, and the aliens that the government simply cannot admit are here but who are ubiquitous. The government cannot explain it to you, but I am going to. But before I do, think of the military and political consequences.

The afterlife is the cloud, and an out-of-body experience is travel in that cloud. We are uploaded and downloaded from life to death and back to life, and we are alive in each state. We can mentally reach into their cloud and extract information. Anyone who has ever had a premonition has done this naturally.

---

[23] Nikola Tesla, https://www.azquotes.com/quote/1069413.

What happened here, with this science, is that every religion just got shoved into the same basket with aliens and each other. Religions are just different views and narratives of something that treats us the same in the fourth dimension. Death. Sorry but not sorry. This makes mankind better and sets humans up for and as galactic neighbors. The aliens already live here.

With the advances in technology, we need advances in ideology.

Advances in sociology and governance must follow. I have just explained the underlying scientific need for and mission of the Institute for Advanced Consciousness Studies (TIFACS). This organization is currently nonexistent, but it needs to be created.

This chapter tells you that you have a soul, and the government knows, and it's proven. Also, there are aliens in the afterlife.

To clear up Locales I, II, and III.

Locale I is the here and now.

Locale II is where the afterlife exists, but that place is huge.

Locale III is the bubble universe; you can "pop" in and out of other dimensions there.

People will tell you that what you have just read here is BS. They are lying because this tips their cart, but it must be done to help all of us. How can you be the best religion when they are all in the afterlife? How can any race be better than another, when what we are is energy, and we change races and genders life to life? You can help, Homer; you can tell people about this.

This chapter proves you have a soul, and you will live again and again and again. In this chapter, you have learned how to prove this to yourself by learning how to visit heaven, the future, or the past and astrally visit Mr. Burns at his house.

Oh yeah, we also need money to study this stuff.

If you are under thirty-five, it tells you what the future will be like and how to navigate it to be successful, protect your family, and do it.

If you are over thirty-five, you may not be able to respond to this information because it more than challenges your beliefs; it overturns them, resulting in a model that rewrites religious law and changes the structure of how we globally govern ourselves and manage our planet. Based on our new understanding of our reality, we must write a common constitution based on wellness and wealthiness to address the fact that aliens are here, and a united response is what is necessary for our common protection.

Wellness is practicing healthy habits daily to attain better physical and mental health outcomes so that instead of just surviving, you are thriving.

Wealthiness is profiting from providing services and conditions to attain better physical and mental outcomes. This goes to the implementation of those laws to achieve this. There are huge business opportunities

The Gateway Process: The Science of the Soul

in this science to those businessmen who understand its significance.

All this can be reduced to one thought: The amount of data that we can manage will soon surpass all the data that comprise our bodies and our lives. Reincarnation is just another way to manage that data. We do it now on our computers and send data loads to the cloud and a new computer; soon data management and reincarnation will simply be a new discipline inside the science of reincarnation. By 2050 we are projected to be able to upload to the cloud and back down all the data that comprises the human body. The evidence that this is already happening is now so overwhelming that funds are needed to explore human consciousness beyond death in a meaningful scientific way. That means mapping the landscape and the time scape.

If time can be likened to a stream, these maps will help you navigate as you go downstream in your life. You can best navigate the stream if you can manage the current. What you are is information.

One final word about how you will react to this information: If you are from the Greatest Generation, born before 1945, you are most likely dead.

If you are a boomer, born between 1946 and 1964, your belief system is so entrenched that this information will be difficult to believe, let alone act on. The likelihood of your response to this information will be resistance. Do not stand aside and benefit from its arrival.

If you are Generation X, born between 1964 and 1981, the college graduates of this generation will see, rather than a belief system, a data management system.

If you are a millennial, born between 1981 and 1998, you will see this information as environmentally sound with good social values.

If you are Generation Z, born between 1998 and 2012, you will see in the upload and download of information to the cloud as the fractal similarity to birth and death. You will see that we are nothing but data coded in our DNA and the quantum energy that produces that state. You will also understand that in your immediate future, we will be able to upload and download more information to the cloud than is contained in the human body.

Chapter 4

The Afterlife

I promised Homer Simpson would understand everything, and so I want to drop my first caveat to Homer. Homer, at some points, the math at the end of this chapter can get intense; it is to prove to bright people that this is our reality. For you, the outcome is the same. You are more than your physical body, you are experiencing multiple lifetimes in various genders, and when you are done, all those experiences will make you more than you are right now. In that explanation lies the journey you are on, so let us go beyond your death to understand why what I just said is true.

This explanation of the afterlife is derived from exit interviews with those who have had a past-life regression or a near-death experience (NDE) and with children who remembered not only prior lives but also their experiences between lives.

Dr. Raymond Moody, who pioneered research into NDE, reportedly said, "How can all these people who don't know each other tell me the same lie?"

The math used in evaluating events is using odds against chance probabilities and looking for fractal patterns in the data. What emerges is the most probable narrative of the afterlife we will present here. It is better and more factual than you have ever heard from a religious cleric.

We have determined that reincarnation has been proven scientifically based on a preponderance of the evidence and the underlying mathematics. Using odds against chance and then looking for fractal patterns, we looked for the most probable scenario.

Reincarnation, however, is one modality of consciousness, and this document maps your consciousness across your own death and the journey you are on. In that sense it provides you a new perspective on your life and what is to come. Whatever religion you are or whatever you believe in, this science validates your beliefs and updates them, providing context and factual information.

You may not believe this premise. The only way to know is to go on the journey this chapter provides and examine the facts for yourself. Then make your own determination.

To understand consciousness and how it operates, we must go beyond your death and back again, and we must prove it beyond a reasonable doubt. In short, you must be convinced this is the case. That will be done with science and math. It is simply the management of the data that is you.

So, what is the common narrative of all these groups? The following is the essence of thousands of people describing what they saw and experienced. Each snippet was woven into this total picture.

You have just died. What happens next?

Proceed to the afterlife after your death.

Arrival Afterlife Cloud

So, you are dead, but because you believed in your religion, you are aware and in what you would call heaven.

That is your belief. But what are the facts as we can see them? Are there things that we have studied or observations that we have that can tell us about what happens next? There are.

That you are just information—your memory, your morals, your situation, and how you feel. All the religions say the same thing and call it a soul. This is all structured information within you and within the cloud. You are, after all, nothing but information. So your specific piece of information meets…

A being of light, an energy orb like yourself, at this point. While there are myriad religious images showing angels and light, this description comes from people who have died and been resuscitated. These could be considered eyewitness accounts, but real scientists would quickly dismiss that type of thinking. But how would real scientists look at all the information? What would the collective look of not just life after death but also consciousness look like? What do you think?

The Applications of the Science of Reincarnation

We are met by a being of light. You may or may not meet relatives. After our initial arrival, we are brought to the processing center. This is what you will encounter: the administration and processing procedure of the afterlife, through which you will pass if you choose to be reborn or, if you will, reincarnated. The indication from all sources seems to be that you can choose to be reincarnated or not. Let us assume you want to be reincarnated. Here is the process.[24]

1. Departure from Earth after your death
2. Homecoming: meeting people you know in the afterlife
3. Orientation: adjusting to a condition of nonlocal consciousness
4. Transition: arriving at your home group of souls
5. Placement: six levels of souls indicate advancement
6. Life selection: choosing your new life on earth
7. Choosing a new body: not just gender but also health and condition
8. Preparation for embarkation: connecting with a group of souls you will be traveling with
9. Rebirth

You are going to follow this path from your death to your next life in a moment. First, I want to explain the

---

[24] Michael Newton, *Journey of Souls: Case Studies of Life Between Lives* (Woodbury, MN: Llewellyn Publications, 2003), Contents.

sources from which this information is taken. Then we will continue your journey. This information was not culled from religious literature but rather from scientific observations.

There were three categories used. The first is NDE, with the original research done at the University of Nevada, Las Vegas (UNLV), by Moody. The second involves children who remember prior lives (CWRPL), done at the University of Virginia (UVA), first by Stevenson and then Tucker. The third is past-life regression, done at Miami by Weiss. Also included in the data was Michael Newton, but caveats are attached to his work. You could dismiss these stories as they have been in the past, but as information processing has grown, we no longer can dismiss the math. The odds against chance calculations we use to determine our reality, from the sun coming up tomorrow morning to batting averages to stock prices, tell us this is our reality. Let us explain what these things are and give examples of each.

For NDEs, take the case of Mellen-Thomas Benedict. In 1982 Benedict died and for nearly two hours was monitored with no vital signs. The story of his afterlife experience is complete and matches other NDE stories, but his particularly, because he was "dead" for so long, contains more information. Early on this type of story was easy to dismiss, even though it supported the concept of an afterlife and heaven, but it seems there are hundreds of millions of NDEs, regardless of race or gender, telling the same story. Raymond Moody's cardiologist's question still stands to this day: How could all these people who do not know each other be telling the same lie? So, we have millions of people experiencing what they call a religious or spiritual event when in actuality they are nonlocally conscious.

"Recent well-conducted studies reveal that about 4.2 percent of Americans public has reported a Near Death Experience. The population of America is a bit more than 315 million people. So over 13 million people have reported having an NDE."[25]

Research has revealed that most people do not immediately report an NDE, so that would indicate this number is much higher. If we extrapolate this to the global population of 7.5 billion, we have more than 300 million people experiencing NDEs.

A word about being nonlocally conscious: We all do it and experience it in greater and lesser degrees based on our own appreciation of what it is and how to use it. Another category that experiences nonlocal consciousness is CWRPL.

Take the case of Suzanne Ghanem, who claimed she was Hanan Monsour.

Suzanne Ghanem was able to identify thirteen separate relatives and their relation to her prior incarnation.

---

[25] Stephan Schwartz, "Six Protocols, Neuroscience, and Near Death: An Emerging Paradigm Incorporating Nonlocal Consciousness," *Explore* 11, no. 4 (July/August 2015).

The odds against this happening are exceptionally high, the equivalent of thirteen heads coming up in a row in coin flips.

The higher the odds against chance, the more probable it is this is our reality, even if it is unaccepted within the structure of our understanding. Her chance of doing this as a child is off the charts.

When Hanan was twenty, she married Farouk Monsour, a member of a well-to-do Lebanese family. Ten days after Hanan died, Suzanne Ghanem was born. Suzanne identified and named thirteen past-life family members. By the time she was two, she had mentioned the names of her other children, her husband, Farouk, and the names of her parents and her brothers from the previous lifetime—thirteen names in all. In trying to locate Suzanne's past-life family, acquaintances of the Ghanems made inquiries in the town where the Monsours lived. When they heard about the case, the Monsours visited Suzanne. The Monsours were initially skeptical about the girl's claims. They became believers when Suzanne identified all of Hanan's relatives, picking them out and naming them accurately. Suzanne also knew that Hanan had given her jewels to her brother Hercule in Virginia before her heart surgery and that Hanan instructed her brother to divide the jewelry among her daughters. No one outside the Monsour family knew about the jewels. Before she could read or write, Suzanne scribbled a phone number on a piece of paper. Later, when the family went to the

Monsours' home, they found that the phone number matched the Monsours' number, except that the last two digits were transposed.

As Suzanne, Hanan was reunited with the Monsours, her past-life family, and demonstrated love and affection for Farouk, her past-life husband. For them to accept her as Hanan, Suzanne had not just to identify them but also communicate little-known facts about her prior life.

The odds against the chance that a child between the ages of two and four are simply so high we must accept this as true, even though most people cannot. This is about recalculating what we think about consciousness against all the new things we are learning. This is the type of work being done at the UVA.

For past-life regression, take the case of the actor Glenn Ford, who remembered not only a prior life as a music teacher but also the town in Scotland where he lived and his name, and he went to the town and found the gravestone. The man whose gravestone it was had been a music teacher who lived the life Ford described.

When Ford was approached about making a movie about Dutch psychic Peter Hurkos, he decided to first study the topic. The fifty-four-year-old actor witnessed some demonstrations by Hurkos and interviewed experts, and in December 1975 he underwent three past-life hypnosis sessions during which he described what appeared to be five previous lives he had led. Dr. Maurice Benjamin conducted the experiment before witnesses with a tape recorder running. The hypnotized actor was regressed back to childhood and beyond, and he described what were presumed to be memories of past lives. In the earliest experience, Ford described himself as a bachelor music teacher named Charles Stewart of Elgin, Scotland, who had died in 1892. Stewart had loved horses but had hated his job teaching music to young schoolgirls. While being questioned about his life as Stewart, Ford agreed to demonstrate his musical skills and played passages from Beethoven, Mozart, and Bach. Ford later listened to the tapes of the interview with interested skepticism. He shared Stewart's love of horses and had, since his early years, been considered a natural with the animals. On the other hand, he could not play the piano. His theory was that perhaps Stewart's antipathy to music and love for horses had carried over to him.

Levels of organization in the afterlife, as described by multiple people and situations, are calculated as odds against chance. This means, by that calculation, this is our reality. This is just the first step in our travel through the afterlife.

Those three observations, the things we see, serve as evidence for judging our reality. The common narrative is proof that our consciousness transcends our deaths and there is an afterlife.

Here is the rub: What we are seeing with this idea is a mapping of that conscious afterlife and an emerging scientific understanding of it. To accept this new idea, one must understand why and how. With our emerging knowledge of information processing, a process is emerging that will make us challenge our beliefs and modify our global and individual actions based on this new information and understanding of it.

The way this sorts out mathematically is that the three categories we observe are each 100 percent certain, based on odds-against-chance calculations. They remained scientific anomalies standing alone, and there was no way to connect them until you looked at them collectively as fractal iterations. At that point a pattern began to emerge, connecting all the narratives in a fractal lattice. When *The Fractal Geometry of Nature* by Benoit Mandelbrot was published in 1982, technology made a huge leap forward using the new math to reduce the size of antennae in a phone and increase data transmission. This science is framing that future technological ability. So what we are offering here is a math proof that not only modifies religious beliefs and how we govern ourselves but also opens new cognitive horizons.

Using odds-against-chance set theory and fractals, we create the most mathematically possible model of reality. This math foundation goes directly to coding and data management. We are, at the end of the day, nothing but information, and our individual fractal bits fit into the overall reality.

Here are why fractals are so important mathematically: They prove this model is right. Nature does not make one of anything, so your being alive is one event in a string of lifetimes for you.

You do not need to understand the math to understand this is right. A fractal is an iteration, each of which is self-similar. Leaves, branching patterns in trees, your lungs, and your veins—all these demonstrate this concept. We believe in this afterlife, and now our ability to process information is confirming those beliefs and removing barriers that exist between us and retard our growth, and this propagates into how we govern ourselves and model our future.

Instead of thinking you are a body that has a soul or a soul who inhabits a body, think of yourself as nothing but information. That is all you are, organized information. That information stays organized whether you have a body or not.

The math calculated as odds against chance across these three categories using a meta-analysis and a fractal determination means this is true and a certainty. Collaborative work still needs to be done, but considering other areas where certainty has now been shown, this may not be necessary.

What we find is that when we lay our common religious experiences over our common observations, we get the following points that affect us all equally.

You will live multiple lives. At some point if you wish to stop reincarnating, you can. It is said you learn more and faster if you reincarnate, but it is not a requirement.

When you die, you are not judged but evaluated, and you are part of that evaluation process.

It does not matter what religion you are or if you do not believe in a religion at all. You can change religion from life to life.

It does not matter if you follow the religious laws of a specific religion rigorously or not, and it does not matter if you change religions; what matters is that you are a good person.

A brief aside, Max Planck said, "I regard consciousness as fundamental. I regard matter as derivative from consciousness." If that is the case, it should explain the validity of the transgender experience to J. K. Rowling if she understands science. Do you see how this science changes the political dialogue?

How would this information, if accepted as the truth, change the world right now? Apostasy, blasphemy, and celibacy would have to be written out of religious law by forward-thinking clerics. If you do not think that is possible, the pope has now accepted the fact that we might meet aliens.

In an interview, French journalist Caroline Pigozzi of *Paris Match* brought up NASA's discovery last July of a new planet, Kepler-452 B, which resembles Earth in its dimensions and characteristics, asking whether there could be thinking beings elsewhere in the universe.

"Honestly I wouldn't know how to answer," the pope replied, explaining that while scientific knowledge has until now excluded the possibility of other thinking beings in the universe, "until America was discovered we thought it didn't exist, and instead it existed."

"But in every case I think that we should stick to what the scientists tell us, still aware that the Creator is infinitely greater than our knowledge."[26]

Would you have thought a pope would have said that thirty years ago? Things change; this is about keeping pace with that change.

We have seen that our observations support the religious view that there is an afterlife and that the odds-against-chance calculations say this is a certainty.

To do this we must see if the narrative of our hypothesis is a fractal of our real-life observations. Let's start with transgenders. All you need to know about transgender people is that they exist and are self-determinant. That means they self-identify.

We see transgender people in our present reality, but the narratives of how that conscious awareness occurs are in the narratives of our observations group. It means that stories emerge of souls changing gender in the afterlife, and we see people coming back claiming it was so and having so much information about their past life. It is an anomaly.

We have those odds against chance, the anomaly

---

[26] Elise Harris, "Pope Francis on Aliens," *The Catholic Thing*, July 6, 2023, https://www.thecatholicthing.org/2023/07/06/pope-francis-on-aliens/.

of the information being certain in both our reality and the reality of the fourth dimension, where heaven is. The fact that both realities, the afterlife, and our present have transgender people points to a fractal alignment.

Take the case of Pam Robinson, a black woman who died and was reborn as a white male. Remember, if Planck's dictum is true, this validates transgender people as normal. Consciousness is fundamental; matter is derivative.

I want to give you an example of an observation in consciousness science and its significance. This category or discipline is called children who remember prior lives, or CWRPL. This kind of observation was once considered an anomaly, something science could not explain. But by connecting it to other anomalous observations and disciplines, we can use them all as fractals to create a mathematical proof of our reality.

The following case study is about a white boy who remembers dying in a fire in Chicago as a black woman in 1993. These are reincarnation stories told by children themselves.

On March 16, 1993, at about 4:00 a.m., a faulty space heater ignited a fire in a room on the first floor of the Paxton Hotel in Chicago, Illinois. The blaze quickly spread to the upper floors, where many of the hotel's 130 residents got trapped. At least 14 people were killed, and many more were injured. When flames roared through a transient hotel, one of the people who died was a woman named Pamela Robinson.

A two-year-old named Luke Ruehlman began to speak of a woman named Pam as soon as he began to talk.

By the time he was five, he was more than mentioning her; he was filling in his parents on her life. When questioned who Pam was by his parents, he said he was in a previous life. The issue his parents had to come to grips with was that their child knew intimate details about the fire and Pam's life—details he could not possibly have known. These emerged as Luke told the story over time, and his parents checked the facts.

Luke remembered jumping to Pam's death to avoid the flames.

The TV show *The Ghost Inside My Child* put Luke in touch with Pamela Robinson's family.

His mom, Ericka Ruehlman, got a photo of Pamela and then mixed them with several bogus photos. Luke accurately picked out Pam's photo, although he had never seen her before in his life.

Luke, after the age of five years old, stopped talking about Pam. This is common in CWRPL. In fact, at the UVA, this is a common theme of souls changing genders. The fundamentalists say this is not true, there are only two genders, and they believe in a soul. But here is proof that the soul they believe in, their soul, will change gender as well. So fundamentalist law should be changed to include not just people who show transgenderism, but all transitory movements should be accepted and celebrated as part of the whole our larger selves are. This is an example of science rewriting religious law to the benefit of everyone.

So how does all this life and death upload and download work? Five hundred years ago, we would have said magic or God. Today we say data management.

So how does data management enter the afterlife and belief? We can run experiments to see if prayer works.

Let us explain it this way: You were built from matter called cells; they were made up of atoms, which were built from quantum particles, which are not particles at all but vibrating bits of energy. That energy is organized into you. That energy is a quantum state where both life and consciousness exist in another form. But still you.

Now I want to introduce you to Stephan Schwartz, and we are going to skim his paper "Six Protocols, Neuroscience, and Near Death: An Emerging Paradigm Incorporating Nonlocal Consciousness,"[27]

> THE NEUROSCIENCES A group of disciplines focuses on the local mind: the neuroscience, the physiological mechanics of an organism's consciousness. These scientists are often not interested in nonlocal consciousness and, indeed, may

---

[27] Stephan Schwartz, "Six Protocols, Neuroscience, and Near Death: An Emerging Paradigm Incorporating Nonlocal Consciousness," *Explore* 11, no. 4 (July/August 2015).

believe it could not exist. Yet by pushing forward to the edge of the physical, they have begun to unravel how the nonlocal becomes local in spite of themselves because nonlocal awareness projects itself into the physiology of their consciousness research.

This is because matter/chemistry/biology is built on physics, as I explained before Stephan began.

> QUANTUM BIOLOGY, another new subdiscipline, posits the following: life is a molecular process; molecular processes operate under quantum rules. Thus, life must be a quantum process. Experimental evidence is beginning to accumulate that this quantum view of life processes is correct.

Now all this consciousness science turns on Planck's assertion that consciousness is fundamental and matter is derivative. That means you were conscious before you were born and do not remember; in fact, those memories fade in CWRPL by the time they are seven or even sooner; for some, they never forget. Stephan added,

> The third front exploring Planck's assertion is work that explicitly studies nonlocal consciousness through experimentation. These studies fall basically into two categories: nonlocal perception,

the acquisition of information that could not be known through psychological sense perception and nonlocal perturbation, consciousness directly affecting matter, including therapeutic intention/healing.

Today there are six stabilized parapsychological protocols used in laboratories around the world exploring these two categories of phenomena. Under rigorous double or triple blind, randomized and tightly controlled conditions, each of these six has independently produced six sigma results; six sigma is one in a billion-1,009,976,678-or the 99.9999990699 percentile.

To the nonscientist, this means it is 100 percent certain. Here are six areas we are studying:

Nonlocal Perception Remote Viewing. A double- or triple-blind protocol in which a participant is given a task that can be accomplished only through nonlocal perception, the acquisition of information that could not be known with the normal physiological senses because of shielding by time or space or both. Sitting in a room two thousand miles away, in answer to the question "Please describe the current circumstances and conditions of the target couple," you could not know they were at that moment standing beneath a waterfall in the mountains of Columbia standing next to the water surrounded by greenery, watching two flying parrots. But nonlocal perception can and has provided just such information many thousands of times under conditions that even skeptics have had to acknowledge are impeccable.

Ganzfeld. A protocol similar in intent to remote viewing in which an individual in a state of sensory deprivation provides verifiable information about film clips being shown at another location.

Presentiment. A measurable psychophysical response that occurs before actual stimulation, such as the dilation of a participant's pupils while staring at monitor screen before the pictures appears. Or, it is a change in brain function before a noise is heard.

Retrocognition/precognition. Many protocols also involve time dislocation to the past or future to be successful. It is routine today to do remote viewing experiments in which the session data are collected and judged against a randomly chosen target set before the target in that set is randomly selected.

Nonlocal Perturbation Random event/number generator (REG/RNG) influence. The REG protocol is actually two major protocols. The first constitutes studies in labs where an individual intends to affect the performance of a physical system, such as a random number generator.

CONCLUSION—TWO MODELS: At present, models of consciousness can essentially be subdivided into two distinct broad categories. Models of the first type: physicalist models holding all consciousness as being contained within the organism's neuroanatomy. Models of the second type: nonlocal models—historically conceived of as esoteric/spiritual/or religious, and distinguished by the assumption that a significant aspect of consciousness is not limited to the neuroanatomy, hence nonlocal.

This implies an awareness and intelligence and self-reference after death, in short what the religious would call a soul. That soul is composed of ordered information. As we have advanced through the last century, we have changed how we handle information.

In short, we will be able to upload and download ourselves to the cloud. What the religions want you to believe and the technology and emerging model of consciousness would support is that this is happening to us already.

In each case, it is the transfer of information. Our technology constantly changes our belief system. Our science is proving a larger understanding of the universe and where and how we fit in.

Whether you believe in a religion or not, this model

of death and rebirth, or upload and download, is simply an information transfer. In fact, if you believe in a religion, this provides proof of your belief. If we look at this model, not from the perspective of our parents but ourselves, we must realize that we are increasing the amount of information we can both process and digitize.

Regarding the development of AI, by 2020 IBM is promising us a brain in a box. The difference between human and artificial intelligence is that on the human side, we measure synaptic operations per second (SOPS), while on the AI side these transactions are measured as floating-point operations per second (FLOPS). That means to create a brain in a box, we will have artificially created something that is two liters in size, has one kilowatt of power, and can do ten million transactions per second, the same wattage and processing power of a human brain.

By 2035 artificial intelligence will be smarter than humans. When AI is smarter than humans, what will it believe? How can we design a belief system for AI without analyzing our own belief system? Will it choose to be a Muslim? A Hindu? A Christian? When AI is smarter than us, will it have a religious belief system at all? Will this emerging scientific model be it?

Therein is the social problem to be studied and addressed. How will an intelligence smarter than us, within the lifetime of everyone living who is fifty or under, affect our society, and how do we manage that change?

Funding in research is our only defense if we wish to grow and protect ourselves from energy threats. So how do you fund that research so that it is collective, cohesive, and beneficial to all? We are going to present a plan to do just that; it is inclusive, logical, and analytic and relies on cold metrics and an intent to produce health and wellness of humanity.

The information that we can store and transfer is increasing. The point at which we can digitize the amount of information contained in the human mind is being measured.

But the model we have advanced indicates that this is already happening to us, and our development in data processing is a fractal expression of our reality. That means there is a heaven, as we believe, but its structure, as revealed by our observations and experiences, is more homogenous than our separated belief system would indicate.

The outer ring points to a nonlocal consciousness that operates for all of us. It explains the efficacy of prayer, clairvoyance, and a consciousness of life after death.

The inner ring is our reality as represented by what we believe and what we see.

So where is the mind then? Stephan stated,

> Recent evidence suggests that a variety of organisms may harness some of the unique features of quantum mechanics to gain a biological advantage.

These features go beyond trivial quantum effects and may include harnessing quantum coherence on physiologically important timescales.

This work is of enormous importance because it is building step-by-step to the most refined quantum physicality. But even its most ardent exponents recognize that it has not given us the fullness of the mind. It has not answered what CU Smith of the Vision Sciences Laboratory at Aston University calls the "hard problem"—the neural correlates of consciousness (NCC).

The neural correlates of consciousness may not be physiological after all, but rather the energy measured in our bodies by devices such as the EKG and MRI. That energy exists in our bodies in structures called microtubules, and George Hammond, in the following excerpts, explains how we theorize microtubules cause life after death. What this simply means is that we are measuring the energy in the microtubules and are deconstructing its code like the code we write and send on a trip to be stored in the cloud.

With this energy you are alive, you are present; without it you are dead and gone. But the energy is still you and present in an eleven-dimension reality after you are dead, and it returns to confine itself to three spatial dimensions and one temporal dimension in your next life.

What we are is our own code.

It is theorized that code is carried in the energy of the microtubules in your body. It fuses biology and physics within your consciousness.

Dreams are simply daydreaming while asleep. Dreams, however, are visual hallucinations as opposed to mere visual recall. This profound difference is caused by sleep, the fact that we are in an altered state of consciousness while asleep. Life after death is the same thing…microtubule-dreaming is a far different state of consciousness than ordinary waking… and in this case it is a full-blown (all five senses) hallucinatory world called heaven. This phenomenological similarity between dreaming and life after death is remarkable…and is certainly the original historical argument for the existence of life after death. Quite recently, a "second line of argument" has now become the existence of "microtubule consciousness" so-called…and its super phenomenal characteristics, which makes such a thing as life after death scientifically plausible! For those not familiar with microtubules, let me brief you. It's been recently discovered that neurons are full of microscopic hollow tubes called microtubules, and light travels through them and memory is recorded in them by light.

Microtubules remain alive for thirty minutes after death, and I believe that life after death is caused by these optical signals. Light is so fast it can download a three-year (proper time) afterlife in a split second, just like a computer downloads a three-hour movie in a few seconds. The uT system must store neuronal sequences; otherwise, you would not be able to remember anything. Ordinarily, these memories are "recalled" at the uT level and played back slowly by the microtubule system at neuronal speed, but it could play them back at full microtubule speed in a special channel if, say, the neuronal system that normally receives them was discovered to be dead (flatlined)! So this must be what the microtubule life after death channel does...it plays back three years of 100 Hz neuronal memory at microtubule speed, 1015 Hz, in 0.1 millisecond...a trillion-to-one speedup, but in this case the neuronal system is no longer "you, the observer," the microtubule system is! Therefore, the observer's "proper time" is the same for this new "uT observer"! So while the bedside observer sees the person die in a few seconds, the dearly departed would spend three years in Heaven (his wristwatch time) during the same instant. While this may seem amazing to nonscientific readers, it is routine to relativity physicists who experimentally see it in atomic physics experiments every day. So, this microtubule system is so fast, it can actually beat death! Any kind of death...even a lightning bolt!

Finally, the all-important argument for this theory is that while the neuronal system cannot produce a "glorified body," the microtubule system can. The reason for this, I believe, is because the existing neurons that would have been connected to the "missing growth deficit body cells" are not capable of actually "firing," so they never appear in a dream (the former fact is evidenced by the rare but well-known accidental firing of these unconnected neurons causing a congenitally missing phantom limb to suddenly appear, full-size). So while we never achieve a glorified body in a dream, I believe the microtubule system inside these normally "inactive growth deficit neurons" is largely functional so that the microtubule system as a whole does possess a "glorified body" (a phantom fully grown body similar to a phantom limb), but quite obviously the only place it could appear is in the afterlife, since it is exclusively a microtubular phantom. [This is the scientific proof of Robert Monroe's second body and the scientific proof of a soul that leaves the body on death.] It appears to me, then, entirely scientifically plausible that we do end life as angels in a spiritual world! Therefore, I conclude from this startling new evidence that life after death is today scientifically very plausible!

George Hammond, Hyannis, Sep. 30, 2017[28]

---

[28] George Hammond, "A Simple Car-Airbag Model of Life After Death,"

What we are seeing at work here is called a fractal process. We see fractals in the branching patterns of lungs and trees; we also see fractals in processes like the upload and download of information to the cloud and the upload and download of our information to the afterlife. Our model now looks like this with the energy from the microtubules.

This energy is you, this nonlocal energy. By becoming alive it/you limit your senses so you can taste coffee and feel sexy. These feelings do not exist in the way they do here in an eleven-dimension space where you are the energy that you call your soul. This is what this scientific proof is; it is hypothetical, but it is the most clear and concise explanation of our perceived reality, backed up by the mathematics that underlies our facts. Around this truth, we posture from our own cognitive bias.

This all comes together in a paper published recently that describes how the quantum soul operates—theoretically. The following is from the abstract:

> The concept of consciousness existing outside the body (e.g. near-death and out-of-body experiences, *NDE/OBEs,* or after death, indicative of a "soul") is a staple of religious traditions but shunned by conventional science because of an apparent lack of rational explanation. However

---
2017, https://www.academia.edu/44527322.

conventional science based entirely on classical physics cannot account for normal in-the-brain consciousness. The Penrose-Hameroff "Orch OR" model is a quantum approach to consciousness, connecting brain processes (microtubule quantum computations inside neurons) to fluctuations in fundamental spacetime geometry, the fine-scale structure of the universe. Recent evidence for significant quantum coherence in warm biological systems, scale-free dynamics and end-of-life brain activity support the notion of a quantum basis for consciousness which could conceivably exist independent of biology in various scalar planes in spacetime geometry. Sir Roger Penrose does not necessarily endorse such proposals which relate to his ideas in physics. Based on Orch OR, we offer a scientific hypothesis for a "quantum soul."[29]

Our model would have to add the component of the universe within which we operate and your consciousness built upon all the operations and scientific conditions we have described.

---

[29] Stuart Hameroff and Deepak Chopra, "The 'Quantum Soul': A Scientific Hypothesis," in *Exploring Frontiers of the Mind-Brain Relationship*, ed. Alexander Moreira-Almeida and Frankling Santana Santos, *Mindfulness in Behavioral Health*, ed. Nirbhay N. Singh (New York: Springer, 2012), https://galileocommission.org/the-quantum-soul-a-scientific-hypothesis-hameroff-chopra-2012/.

There are two conclusions to this information. First, the mathematical model presented on the science is incontrovertible as being the best and most likely representation of our reality. How you react to this and the depth of your understanding depend on the generation you are from. There is no set of alternative facts that even comes close to incorporating how consciousness functions within the materialist dimensions, as opposed to the quantum dimensions, and then presenting itself as a unified model.

This information should change society in the coming generations, beginning now. We are all affected equally. With the impending data improvements coming from AI, which is predicted to be smarter than us within twenty years, and clustered regularly interspaced short palindromic repeats (CRISPR), which is modifying us genetically, humanity must find a way to incorporate the logic of AI. What must change is governance. Not who runs governments. For the model to work, it must be win-win because when global trouble arrives, it affects us all, and all of us are part of the solution.

Religious law must be rewritten globally by forward-thinking clerics removing such sanctions as apostasy, blasphemy, and celibacy. If the pope can acknowledge the possibility of aliens, we can all acknowledge leaving a religion is no cause for a death sentence.

Finally, a common constitution needs to be written based on wellness and wealthiness. Politicians globally can look to this common constitution to take articles from it and import those articles into their own constitutions, knowing that scientifically, the way the articles of the common constitution are written is the most optimal way to write a constitution to produce a healthy society. At its core the original French bill of rights should look very much like individual guarantees given to every individual by the global banking system. The global economy would be robust given this case, and humanity would truly be able to leave Earth.

We are saying you are the energy, not the body you currently wear. This information changes how we worship and rewrite our religious laws, throwing out our apostasy, blasphemy, and celibacy religious law. It changes how we govern ourselves by writing a constitution based on wellness, which countries can adopt portions of to the scientific and factual benefit of all citizens, rich and poor alike, because all should be engaged in this noblest of human work, the exploration of ourselves.

As we acknowledge and explain this new nonlocal dimension, we meet the nonbiological entities that Lieutenant Colonel McDonnell warned us about.

Do we go forward fighting and killing ourselves?

Before we can successfully do that, we need to deal with our own politics. Our planet is dying, and we go on, as a species, killing each other. For those who know (not you, Homer), we have already had disclosure, but no one knows what to do about it.

From the alien side, we are a useful stop when traveling the galaxy, and why would aliens give us the technology to the stars when we are a bunch of violent assholes? That is a fair question.

In adopting this science as our reality, which, if you are a fundamentalist, you already believe, then why the killing?

Fundamentalists of all types need to stay in their own lane and accept all others regardless of gender, gender presentation, color, race, or belief.

This acceptance protects you; your heaven is a small part of the greater universe, the afterlife, and one small part of a multidimensional universe. All souls of like energy can congregate in the afterlife, but energy travels in its own direction not bound by dogma. The common man's understanding of dimensionality changes the global political perspective.

This cannot be done by the present generation, and this explanation is not aimed at them but at those younger who can see this as an information matrix, and dogmas presented anywhere are blocks to flow. When this is unblocked and kind intent is exhibited, alien contact will expand. We will explain how that is happening now and how anyone can have the experience.

So, we present a united front to these alien threats and establish an optimal to face this new landscape. Kindness.

To do that we need to fund research.

For scientists to openly say this would mean the loss of their jobs.

Chapter 5

Humanity's Current Situation: It Is More Complex Than You Think

As unbelievable as the information in this chapter is to the casual reader, the following presentation is the centerline of truth. This may be easier for Homer to believe than for the average American citizen. The question to be asked of any reader of this chapter is, what would you have done if you had been in charge, and given that this is the current situation, what would you do now?

There is so much misinformation and disinformation when dealing with information these days that we must use the following method:

The Four Questions:

What do we know?
How do we know what we know?
What does it mean?
How is it applied?

This chapter is the core of our situation. Early in my academic career, I was told that if you want to know what the truth is, you must read the extremist views on either

The Applications of the Science of Reincarnation

side of an issue. That will help you find the centerline. So here, in the simplest of terms, we will explain your personal relationship to the problems that we face as a race on Earth. We are visited by multiple species of extraterrestrials and transdimensionals, and sometimes they are both.

For humanity as a species to understand our position in this universe, technology transfer is not enough; there must be a common consciousness understanding that removes the blocks that disunite us because that common consciousness has an effect in a transdimensional world. We occupy a transdimensional world, and most of us do not know it. As a species, we are becoming transactional in that environment, and in consciousness and the science of reincarnation, we find the hinge on which our future turns.

For beginners, the science of reincarnation has mathematically proven that you will not die and that your personal mind has an address in space and time. You are not a body that has a soul; you are a soul here to have a human experience, and you shed your body through your life and after death. All the manifestations of psychics, remote viewers, and clairvoyants are transdimensional experiences, as well as the ones where aliens communicate telepathically with each other and with you. You are transdimensional, and most of us do not yet know that. So rebirth, in a sense, is being reborn in a transdimensional environment or being uploaded and downloaded to your particular cloud of choice (insert your religious belief here).

The science of reincarnation is not just individual; it is societal as well. This science goes to governing and self-management.

In explaining the science of reincarnation to the layman, it must be said that you, my reader, are just data. You already know you can upload and download information to a new computer. We will meet a race of aliens, the Grays, that are using this and humanity to create a hybrid race to move their "essence" to. In reality, it is just a transfer of living information that differentiates uploads from downloads. This goes hand in hand with nonlocal consciousness and intelligence gathering.

I want to be clear about my position regarding the following choices made by our government on our behalf in the name of the United States National Security. As you read it, ask yourself what you would do or what you would have done if you were involved as the events unfolded. We only need to look at the response to the Orson Welles radio broadcast only decades before that created mass hysteria.

Our relationship with aliens changed in 1946 with our entry into the Atomic Age.

I am not going to try to make the case that aliens are here; the proof is now incontrovertible. I am going to try to deal with what should be done about it, how it is being handled, and the consequences of what is occurring.

ALIENS—the turning point. The event at Roswell is not the beginning of alien interference and management of the human race but marks a turning. At the time of the Roswell crash, Roswell was the only nuclear-armed station in the world.

The alien crash resulted in the acquisition of materials that could be studied and created a sea of change in the whole national security organization within weeks.

The Roswell crash occurred on July 8, 1947. Roswell Army Air Field issued a press release stating that they had recovered a "flying disc." The army quickly retracted the statement and said instead that the crashed object was a conventional weather balloon. The United States had positive proof, including ships, alien bodies, and one alien still alive.

Two months later, on **September 18, 1947**, the Central Intelligence Agency (CIA) was formed when Harry S. Truman signed the National Security Act of 1947.

On **September 26, 1947**, by order of the Secretary of Defense, the Army Air Forces (AAF) were transferred from the Department of the Army (formerly the War Department) to the newly established Department of the Air Force, officially creating the United States Air Force (USAF).

According to historical records, the creation of "Unacknowledged Special Access Projects" (SAPs) is generally attributed to the passage of the National Security Act of 1947 on July 26, 1947, which established the framework for highly classified intelligence programs within the US government, allowing for the development of projects that could be kept secret from most government officials and the public.[30] This is the document we referred to at the end of chapter 3. A question for my American readers: How hard is it to get rid of an entitlement program once instituted? How about one that has in its charter that it is granted total secrecy?

Nellis Air Force Base is where this newly acquired alien material ended up. It is located in Las Vegas, Nevada, and is about five thousand square miles in size. It is located more than 120 miles northwest of Las Vegas.

Beginning in 1947, Nellis Air Force Base became the epicenter of alien technology research in the United States. The Area 51 addition was even more remote and secure, and it was here that the United States began to try to retro-engineer the vehicles they had recovered.

The US government's official name for Area 51 is the Nevada Test and Training Range, which is a unit of the Nellis Air Force Base. Today it is used as an open training range for the US Air Force.

According to the CIA, the name Area 51 comes from its map designation. It was also previously referred to as "Paradise Ranch" to make the facility sound more

---

[30] Response to "Date of Creation of Unacknowledged Special Access Projects," Google AI Overview. 9/30/24

The Applications of the Science of Reincarnation

attractive to those who would be working there. "Paradise Ranch" was then shortened to "the Ranch."

It is restricted to the public and has armed guards patrolling the perimeter. It is also impossible to enter the airspace above without permission from air traffic control.[31]

How do we know what we know?

I am now going to use Dr. Michael Salla's book, *Exopolitics: Political Implications of the Extraterrestrial Presence*, regarding the timeline of events. It is well-sourced, and I do not need to recreate what Dr. Salla has explained and documented. In explaining the events as they unfolded, we can see governmental policy taking shape.

The restrictions imposed by the intelligence community on the information limited their ability to respond to it. While governmental bifurcation may have been appropriate at the time as events unfolded, over the next seventy-five years, it restricted the government's ability to bring resources to bear on the problem.

My job here is not to reiterate what Dr. Salla lays out but to respond to the information scientifically, in innovative, beneficial, and intelligent ways. To do that, complexity must be taken out of the equation because it obfuscates a larger, more important truth. What do we do about the situation we are in?

---

[31] "Area 51 Fast Facts," US Crime and Justice, CNN, updated August 7, 2024, https://www.cnn.com/2019/07/31/us/area-51-fast-facts/index.html.

This is fair use of copyright, as Dr. Salla is spot-on with his observations, so I will use Dr. Salla's well-sourced timeline so we may deconstruct both policy and reaction on all sides of the political and intergalactic landscape. I want to point out the difference in policies between extraterrestrials and transdimensionals as part of creating an exopolitical platform for humanity. Exopolitics is the political approach to interactions beyond Earth, while Earth politics happens on Earth. Exopolitics happens off Earth.

I will cover only a few alien groups, as there are too many to include in a complete anthology here. Remember, each of these alien groups can be reached nonlocally by the remote viewing group that it should work in conjunction with. The naval policy of "Look before you leap" can be implemented here as well and will be discussed in the chapter "Army Futures Command."

It would be counterproductive to this proposal to attempt to recreate the arguments that aliens are here, but we will take the centerline of information and give examples of where it has been extracted from.

Eyewitnesses

Aside from Buzz Aldrin, Edgar Mitchell, millions of witnesses, whistleblowers, and abductees, you have a parade of presidents alluding to aliens being here as well, providing a wealth of sources.

As we learn about the galactic neighborhood, stories emerge.

Remote Viewing

"The information on ETs gathered by remote viewers is often startling in its implications. For instance, Brown describes remnants of a Martian civilization and its relationship with Earth: 'There are Martians on Earth, but one must think clearly about the implications of this before ringing the alarm bell. These Martians are desperate. Apparently, they have very crude living quarters on Mars. They cannot live on the surface. Their children have no future in their home world. Their home is destroyed; it is a planet of dust."[32]

Perhaps more significantly, remote viewers described a federation of worlds responsible for controlling evolving planets such as Earth. In various sessions, remote viewers described the nature of the headquarters of this Federation and the ways these ET races influenced behavior.

Abduction

"But what I have found to be so extraordinary from the beginning of my study has been the readily identifiable

---

[32] Michael E. Salla, *Exopolitics: Political Implications of the Extraterrestrial Presence* (Tempe, AZ: Dandelion Books, 2004), 16.

patterns that emerge when the case narratives are examined carefully…The growing number of abduction cases has led to this category rising in importance as a credible source of evidence for an ET presence. Estimates of the number of abductions in the US go as high as several million, signifying a huge database of potential information on the ET presence. This database is continuously being refined as researchers develop the analytical…"[33]

The point of the last quote is to reference what Raymond Moody says about near-death experiencers. The readily identifiable themes lead to an analysis of this being our reality. The math proof that this is occurring is now a certainty. This same math proof proves the transcendent nature of consciousness we all, aliens, and us, are involved in. So, we are not just dealing with extraterrestrials; to cover the huge distances of space, they need transdimensional technology. The frame of reference my reader must use to grasp the landscape is that their view of life after death just became their reality, but so did everyone else's.

"Some former officials and/or employees involved in clandestine government military projects report to have seen live ETs at various military bases. Michael Wolf, a former member of a National Security Council committee that oversaw the ET presence, claimed that he 'met

---

[33] Michael E. Salla, *Exopolitics: Political Implications of the Extraterrestrial Presence* (Tempe, AZ: Dandelion Books, 2004).

with extraterrestrial individuals every day in my work, and shared living quarters with them.' The reports of joint ET-human bases give credence to reports that a treaty was signed as early as 1954, during the Eisenhower administration, granting rights to ETs to establish joint bases for technological sharing on US soil and placing strict limits on the biological experimentation the Grays were conducting with US citizens. The existence of such a treaty between the Eisenhower administration and an ET race has been disclosed by a number of government/military whistleblowers."[34]

William Cooper, part of a naval intelligence briefing team for the commander of the US Pacific Fleet, narrates the events surrounding this treaty from classified documents he claims to have read as part of his official duties:

"In 1954 the race of large nosed Gray Aliens which had been orbiting the Earth landed at Holloman Air Force Base. A basic agreement was reached. This race identified itself as originating from a planet around a red star in the Constellation of Orion which we called Betelgeuse. They stated that their planet was dying and that at some unknown future time, they would no longer be able to survive there. This led to a second landing at Edwards Air Force Base. The historical event had been planned in advance and details of the treaty had been agreed upon.

---

[34] Michael E. Salla, *Exopolitics: Political Implications of the Extraterrestrial Presence* (Tempe, AZ: Dandelion Books, 2004).

Eisenhower arranged to be in Palm Springs on vacation. On the appointed day the President was spirited away to the base and the excuse was given to the press that he was visiting a dentist. President Eisenhower met with the aliens and a formal treaty between the Alien Nation and the United States of America was signed."[35]

"Alluding to a treaty signed by the Eisenhower administration, Col. Phillip Corso, a highly decorated officer who served in Eisenhower's National Security Council, wrote: 'We had negotiated a kind of surrender with them as long as we could not fight them. They dictated the terms because they knew what we most feared was disclosure.' Another whistleblower to elaborate on this treaty was Phil Schneider, a civil engineer who had worked on the construction of secret underground facilities, and who was found dead seven months after giving a lecture in May 1995 with the following information: Back in 1954, under the Eisenhower administration, the federal government decided to circumvent the Constitution of the United States and form a treaty with alien entities. It was called the 1954 Greada Treaty, which basically made the agreement that the aliens involved could take a few cows and test their implanting techniques on a few human beings, but that they had to give details about the people involved. Slowly, the aliens

---

[35] Michael E. Salla, *Exopolitics: Political Implications of the Extraterrestrial Presence* (Tempe, AZ: Dandelion Books, 2004), 157–158.

The Applications of the Science of Reincarnation

altered the bargain until they decided they wouldn't abide by it at all...

"Former officials involved in clandestine projects run by different government agencies, claim that the Grays communicate through telepathic thought exchange, and use thought to activate the advanced technology and navigation instruments of their spacecraft."[36]

"In reports on the content of the telepathic communication with the Grays, abductees/contactees have reported that the Grays' survival as a species is threatened due to genetic degradation as a result of the repeated use of cloning for reproductive purposes.

"The Grays say they need to create a hybrid race that integrates the human and Gray genes in order for the continuation of their species. It is claimed that with the creation of this hybrid race, the Grays will somehow be able to transfer their "essence" or "consciousness as individuals into the hybrid, thereby perpetuating the continuation of their race."[37] Read this as the science of reincarnation as used by other species. When they transfer their essence, it means who they are. They migrate to the new body; the same things humans do when dying and being reborn or when they shed their bodies as they grow. No one has the same body they had when they were

---

[36] Michael E. Salla, *Exopolitics: Political Implications of the Extraterrestrial Presence* (Tempe, AZ: Dandelion Books, 2004), 23–24.

[37] Michael E. Salla, *Exopolitics: Political Implications of the Extraterrestrial Presence* (Tempe, AZ: Dandelion Books, 2004), 24–25.

six, but they have memories of that time. Consciousness in all creatures is migratory across their life/lives.

There will be readers who pass this point and do not believe in the science of reincarnation, but this is not about belief. *This is a threat here on Earth that, without anyone's help, humans may annihilate themselves.* But to the point, do not let your prior beliefs limit what you see. In a dimensional manifold, the science of reincarnation is very real, both in an individual sense and a collective sense. The organization we remote-view, which helps developing worlds like ours, sees Earth as one organism. Their goal seems to be to have Earth be an intelligent one.

There are several points to make regarding alien use of the science of reincarnation. Consciousness science is not just about human consciousness or alien technology but also about retro-engineering alien consciousness and developing our own technology for consciousness uploads and downloads.

On the science of reincarnation, scientists here must embrace the entire concept. However, the inhibitions from an entrenched worldview and power structures are immense. This science is directly tied to LGBTQ policy and religious nationalism.

Once you acknowledge the alien presence and adjust to how the new landscape shapes up, we find these Grays do not pose an overtly hostile threat. Even though we have tried to attack them, there have been few, if any, hostile actions taken against us. The United States has tried to attack alien ships in our airspace deliberately. The aliens have not responded; they simply avoid us. They have better technology and presumably better weapons, and despite our attempts to attack them, they have not attacked us. Additionally, while abduction is frightening, the people taken are not harmed and are returned. The estimates, as stated, run in the millions.

According to Dr. Salla,

> The idea that telepathy is the standard communication means of ETs gives plausibility to the claims of numerous civilian sources that they have been contacted by ETs and conducted telepathic communication with them. This is especially important information given by "channelers" who claim to communicate with different ET races. In sum, the "intruder perspective" describes the intentions, abduction activities, genetic experiments, violations of the United States and allied airspace, and advanced technology of ETs as intrusive. Moral categories used to describe the Gray intruders are unclear.[38]

Ingo Swann talks about a telepathic language that humans do not yet speak. We learned in chapter 3 that we can all do this and how to do it. Just buy the Monroe tapes

---

[38] Michael E. Salla, *Exopolitics: Political Implications of the Extraterrestrial Presence* (Tempe, AZ: Dandelion Books, 2004), 25.

and you can learn to talk to aliens on the couch in your home. I caution you, though: Read what experiencers say on Reddit and prepare yourself. Like any sojourn into unknown waters, there lies both danger and opportunity.

We access this field when we remote-view, this field of shared mind, and here we find the Pleiadians. Here's Dr. Salla's definition of Pleiadians:

> "The Pleiadians say Family of Light [humans committed to transformational change] are *systems busters* [emphasis added] who travel through time to systems in need of change, helping to facilitate the collapse of these systems.
>
> Marciniak and most of these "channels" support the idea that an ET presence has existed on the planet for millennia, and that there exists an officially sanctioned government conspiracy that is global in its reach to keep this presence secret from the general public. The testimony of channels in a court of law would obviously be controversial and not likely to be conclusive in any way. Nevertheless, the overwhelming evidence suggesting that ET races do communicate with humans through telepathic exchanges, indicates the importance of not excluding this body of evidence."[39]

---

[39] Michael E. Salla, *Exopolitics: Political Implications of the Extraterrestrial Presence* (Tempe, AZ: Dandelion Books, 2004), 19.

We also find the Anunnaki.

When Courtney Brown remote-views the Anunnaki, he describes the Anunnaki remnants in space around us as often being in conflict with the ET intruders, the Grays. The Anunnaki are sometimes referred to as Reptilian.

According to Dr. Salla,

"They argue that clandestine government agencies are aware of the Anunnaki remnants on Earth and are divided over how to respond to the anticipated return of the Anunnaki elites in the immediate future, that may attempt to reestablish overt control over the planet. Brown, Picknett, and Prince claim that clandestine government programs are using "exotic" technologies such as psychotronic weapons activated by thought, enhanced psychic abilities, "star gates," and even time travel to militarily prepare for the return of the Anunnaki who possess a technology supposed to be superior to anything reverse-engineered from downed Gray spacecraft.

A number of individuals who claim to have participated in clandestine military programs with ET races have also reported interacting with or witnessing a large reptilian species that presumably is associated with the historic Anunnaki presence on the planet. Andy Pero, who claims to have been recruited as a super soldier by a clandestine government organization, gave testimony of such an encounter:

On one occasion I was introduced to a reptilian being while in an underground base sometime in 1989–90. At first I saw a seven-foot tall, human, Aryan-looking man. He walks towards me and I noticed that his image phases out as if something interfered with an energy field. He does something to a device on his belt and tells me, "OK, I'll show you." He then pushes some button, and then I see his image change into a seven-foot-tall lizard-like creature who looked like he weighed over four hundred pounds."[40]

According to other participants, including Michael Relfe, Al Bielak, and Preston Nichols, "Reptilian beings are involved with time/interdimensional travel technology, mind control, psychotronics, and other exotic weapons."[41]

I want to pause and reference Robert Byrd's trip to the Antarctic. In his papers he described meeting a race of more advanced beings who lived at the South Pole, and when his plane was taken into their space and the weather changed, he reported seeing mastodons. He was transdimensionally taken across a border to a South Pole that was

---

[40] Michael E. Salla, *Exopolitics: Political Implications of the Extraterrestrial Presence* (Tempe, AZ: Dandelion Books, 2004), 28–29.

[41] Michael E. Salla, *Exopolitics: Political Implications of the Extraterrestrial Presence* (Tempe, AZ: Dandelion Books, 2004), 29.

warm and temperate. What Salla describes here is cross-confirmed by related stories, building an entire map:

> Other sources report that during the early years of the Eisenhower administration, a group of ETs secretly met with United States government officials appointed to deal with the ET question and offered to assist with a number of environmental, technological, political and socioeconomic problems with the sole condition that the United States dismantle its nuclear arsenal.
>
> When the government officials declined, this group of ETs subsequently withdrew and played no role in the government's clandestine program to reverse engineer ET technology for advanced weaponry. It is claimed that these "helper ETs" would subsequently concentrate their efforts on the consciousness raising of the general public—warning of the hazards of nuclear and "exotic" weapon systems reverse engineered from ET technology and limiting the environmental impact of clandestine projects.
>
> "Then it is clear that the study of exopolitics will be compounded by officially sponsored acts of disinformation, official denial, and even intimidation by clandestine government organizations."[42]

---
[42] Michael E. Salla, *Exopolitics: Political Implications of the Extraterrestrial Presence* (Tempe, AZ: Dandelion Books, 2004), 4.

Now what needs to be done here in an exopolitical sense is to make policy recommendations, which is tough to do in a world of secrecy. Not disclosing this information in a concise form limits resources very badly needed by both our intelligence community and not just the US government but all world governments. The intelligence services need our help (by that, I mean billionaires and scientists) but do not know how to ask.

This chapter lays out the conceptual foundations for developing suitable policy responses to the ET presence that support the need for exopolitics as a field of study.

Now this goes hand in hand with the study of consciousness science.

Governmental Bifurcation

Proof of aliens was the number one secret the United States had that was more important than its nuclear arsenal.

To handle this, special access programs were set up with firewalls between the politicians, the military, and the US government itself. Barry Goldwater once asked Curtis LeMay if he could see what the US government had.

> "On March 28, 1975, Goldwater wrote to Shlomo Arnon: 'The subject of UFOs has interested me for some long time. About ten or twelve years ago I

made an effort to find out what was in the building at Wright-Patterson Air Force Base where the information has been stored that has been collected by the Air Force, and I was understandably denied this request. It is still classified above Top Secret.' Goldwater further wrote that there were rumors the evidence would be released, and that he was 'just as anxious to see this material as you are, and I hope we will not have to wait much longer.' The April 25, 1988, issue of *The New Yorker* carried an interview with Goldwater in which he recounted efforts to gain access to the room. He did so again in a 1994 *Larry King Live* interview, saying:

'I think the government does know. I can't back that up, but I think that at Wright-Patterson field, if you could get into certain places, you'd find out what the Air Force and the government knows about UFOs...I called Curtis LeMay and I said, "General, I know we have a room at Wright-Patterson where you put all this secret stuff. Could I go in there?" I've never heard him get mad, but he got madder than hell at me, cussed me out, and said, "Don't ever ask me that question again!"'[43]

---

[43] Intelligent-Pause510, "Serious Post: Barry Goldwater should have something to do with UFO's," Reddit, 2023, https://www.reddit.com/r/TNOmod/comments/y8iapu/serious_post_barry_goldwater_should_have/.

In 1961 Eisenhower, when he left office, had lost control of this separate and secret organization. What he called the military-industrial complex should be renamed the military-industrial-intelligence complex.

Here are some questions to consider: Did the aliens who wanted us to give up nuclear weapons have nuclear weapons? Are nuclear weapons a threat to dimensional time-space? Their dimensional time-space? Ours?

What does that mean to you, the taxpayer? Deep black projects run over $200 billion per year, and the Department of Defense (DOD) does not account for 25 percent of what it spends. They can, but they just will not. When the DOD says they cannot account for it, it means they will not account for it.

What are we getting for our money?

I do not want to indict our intelligence community; in many instances, they give up their lives for ours, but they are overloaded. So how do we proceed?

These are brave men and women working to protect us all. What would you have done? What would you do now? But how do you bring them back in? How do you support them? Seriously. You see, they now have the technology to change the world. The power source of the alien spacecraft and our retro-engineering of it.

But they are *waaaay* overmatched.

The Congressional Research Service at the Library of Congress Science and Technology Division, which reviews all material, both classified and nonclassified, estimates there are between two and six other highly developed civilizations just in the Milky Way.

So, if we use four as the number between two and six we can use four times that number, assuming a 25 percent success rate. Then there are sixteen as developed as us, meaning that there are twenty civilizations either more developed than or as developed as we are in our galaxy.

So, twenty planets at our level in this model, accessible planets in this galaxy times 125 billion galaxies, as estimated by the Hubble Deep Field Telescope, is 2.5 trillion planets. You have the technology to reach them all and none of the resources in terms of human capital, and with a reductionist model operating here on Earth, you are going to get hurt out there. All one needs to do is look at the condition of the Russian oligarchs. This is a limited view in 3D space.

What is the change in groupthink needed to create acceptance of a 4D model and keep control of the process and the transition? The proposals we are about to make are about this.

The science of reincarnation is the key to opening that lock. It leaves tradition and religion without teeth. It respects the past but looks to the future and, in doing so, puts down arms.

The change will occur in one generation. Getting rid of nuclear weapons is a peripheral political benefit because this requires a global constitution.

If getting rid of nuclear weapons and the pariah of religion is my hall pass to the universe, then the ones who profit from wars have just found a better, more profitable, larger game.

We are not just dealing with aliens; we are also dealing with the common mind, our collective common mind.

This explanation of consciousness needs to evolve because if you use one remote viewer, ten remote viewers are better, and global consciousness is better still, regarding understanding extraterrestrial intelligence and extradimensional intelligence and its exploration.

Thanks to Mike Salla's well-researched timeline, we can simply explain the dynamic gestalt of our situation:

1. There are aliens with many different agendas.
2. There are transdimensionals.
3. We, human consciousness, and religions are all the same. Just dress and custom are different to make people be more submissive and controllable. Continuity of consciousness is a fact for us all.
4. We, human consciousness, all colors are the same. Bill Nye said you enter this world in a band around the globe that dictates skin color. Near the equator is darker, yet near the poles lighter. Differentiating this is another way to keep us apart. But common consciousness seems to be what the federation we remote-view is interested in. Common health can be infected to do harm to the body, and infectious

viral ideas need to be laid to rest by courageous scientists before they infect the mind. Ideas antithetical to logic must be addressed even when, and especially when, it contravenes religious belief.
5. Is the process protected? In short, is the development of civilization protected? Are we being protected from the Anunnaki? If so, by whom? Courtney Brown refers to the federation. He has asked them for help to remote-view the Reptilians who tried to block him. Shouldn't this have some priority for humans?
6. Our government is outmatched. Government, just who do you think is getting into those ships you have? This is going to consume more resources, and to quote a line from the movie *Jaws*, "You're going to need a bigger boat."

Homer, there are aliens all around us, and the only way to protect ourselves is to acknowledge our common consciousness. To do this we need your help to remove false divisions between us. We need your help, Homer; that is scary.

Chapter 6

Not Disclosure but Context

Humanity needs context on disclosure. This book presents a plan that involves us all and provides a soft landing. Without the science of reincarnation, there is no logical way to explain nonbiological entities or dimensional access. For the moment let us stay in this dimension.

To synopsize, we have proven mathematically that our consciousness transcends our permanent bodily death, and there is a continuity of an individual's consciousness after death. We have explained the energy in the microtubules has been identified as your soul. There is still some discussion about the makeup of this energy. This makes us conscious in a new dimensional space, a 4D awareness. This proof scientifically explains clairvoyance, remote viewing, the efficacy of prayer, heaven, and the electromagnetic nature of consciousness. Individually many describe "seeing" this 4D space with their third eye. As a race, though, we are third-eye blind. So, I want to give you a consensus third-eye view of our reality. I must tell you about our most scientific consensus view of aliens/extraterrestrials to do that.

So when will there be disclosure? It has already happened, and it is not going to. That statement explains that, based on what has been disclosed, we already know that aliens are here. The government is not going to admit it because there is no response they could offer that would be expected. That is the reason for this proposal. Governments will only acknowledge what they are forced to, and the information will continue to leak until, like a jigsaw puzzle, we will see the whole picture, even though some pieces are still missing. But some of the pieces are considerably large. One of the best examples is an edited recording of the May 9, 2001, meeting at the National Press Club in Washington, DC, where over twenty military, intelligence, government, corporate, and scientific witnesses discussed the reality of UFOs or extraterrestrial vehicles, extraterrestrial life-forms, and resulting advanced energy and propulsion, hosted by Steven M. Greer.[44]

If the disclosure you seek is governments acknowledging alien contact, you will never hear the truth. You will hear some truth and some lies because of national security concerns, profit to be made from emerging technologies for a select few, and general incompetence at not being able to admit to a reality that will reshape the world.

---

[44] Steven M. Greer, *Disclosure Project: National Press Club news conference*, May 9, 2001, https://search.worldcat.org/title/Disclosure-Project-:-National-Press-Club-news-conference-May-9-2001/oclc/52741633.

Managed Global Contact

Let us explain our current situation by painting a picture of what has been disclosed and by whom, and you will begin to see the gestalt of the problem we all face. If you have a seat belt, buckle it. I will tell you about aliens from an insider's perspective and give you a consensus picture extracted from sources in the public domain. You will know many—all are verifiable—and at the end of the description that these individual bytes will produce collectively, you will have a more cohesive view of the landscape we are working in.

Regarding alien presence, aside from cell phone footage worldwide, the US government has acknowledged this activity in their unclassified report from the Office of the Director of National Intelligence, titled "Preliminary Assessment: Unidentified Aerial Phenomena," dated June 25, 2021.

It makes the following points:

1. "There are probably multiple types of Unidentified Aerial Phenomena (UAP)."
2. "UAP poses a safety of flight issue and may pose a challenge to US national security."
3. "Explaining UAP will require analytic, collection, and resource investment."
4. "The UAPTF (The Unidentified Aerial Phenomena Task Force) is looking for novel ways to increase collection of cluster areas."

This admission was forced by the ubiquitous nature of cell phone footage and the navy's cameras showing technology we, as humans, do not possess.

This admission is peeling back the onion. There has been technology transfer from alien species to humans for years.

So let us run through narratives lifted from the media to get a picture of what is out there.

In an article by Vicky Verma, it is stated, "Former Canadian Minister of Defense Paul Hellyer said that aliens not only exist but are on Earth among people. He said that there had been four alien species visiting Earth for thousands of years."[45]

The article goes on to say; "In support of his words, the ninety-year-old ex-minister spoke about one incident in 1961 when Canada noticed a group of fifty unidentified flying objects moving from the Soviet Union to Europe. Then this group of objects suddenly went to the North Pole and disappeared. Hellyer noted that the investigation of this incident lasted three years, and it unequivocally showed the activity of extraterrestrial civilizations that flew to Earth to ensure that people would not use nuclear weapons.

According to Hellyer, aliens already live on our planet: Most aliens are like humans, so it is almost impossible to notice them in the human mass. He first spoke openly

---

[45] Vicki Verma, "At Least Four Alien Species Have Been Visiting Earth for Thousands of Years from Andromeda," How and Whys, January 21, 2022, https://www.howandwhys.com/four-alien-species/.

about his belief that governments are covering up an alien presence back in 2005, saying UFOs are as plentiful in our sky as airplanes. "Much of the media won't touch it, so you just have to keep working away at it, and we will get a critical mass, and one day they will say, Mr. President or Mr. Prime Minister, we want the truth, and we want it now because it affects our lives."

He said that aliens had visited Earth from different star systems, including the Pleiades and Andromeda. "There are extraterrestrials from Andromeda, and ones that live on one of Saturn's moons. There is a federation of these people, and they have rules; one of them is that they don't interfere with our affairs unless they are invited."

The former Secretary of Defense of Canada noted that aliens are not aggressive and have a huge stock of knowledge. However, guests from other planets do not want to share their knowledge with people yet, because they are afraid that earthlings would use them for wars, and not for good purposes."[46]

In an article from *The New York Post* written by Yaron Steinbuch, we have the following excerpt:

> Former CIA Director R. James Woolsey said he believes UFOs could exist after his friend's plane

---

[46] Vicki Verma, "At Least Four Alien Species Have Been Visiting Earth for Thousands of Years from Andromeda," How and Whys, January 21, 2022, https://www.howandwhys.com/four-alien-species/.

was "paused at 40,000" feet—and hopes humans would be friendly to extraterrestrials if they ever make contact, according to a report...

In December (2021), ex-CIA Director John Brennan said it was "presumptuous and arrogant" to believe there are no other forms of life than the ones on Earth...

In April 2020, the Department of Defense had navy videos from 2004 and 2015, each showing "unidentified aerial phenomena"...

Last month, former Director of Intelligence John Ratcliffe said on Fox News that "there are a lot more sightings than have been made public"...

"And when we talk about sightings, we are talking about objects that have been seen by navy or air force pilots, or have been picked up by satellite imagery that frankly engage in actions that are difficult to explain," Ratcliffe said...

"Movements that are hard to replicate that we don't have the technology for. Or traveling at speeds that exceed the sound barrier without a sonic boom," he added...

The UFOs were seen moving at incredible speeds and performing seemingly impossible aerial maneuvers. One of the clips showed a dark circular object flying far in front of a jet, while a second caught a small object racing over land.

The third captured a circular object first

speeding, then appearing to slow down—and moving closer to the pilot's camera.

The Pentagon announced in September 2021 that it created an Unidentified Aerial Phenomena Task Force...

The latest developments come amid rising interest in UFOs, which has been indicated by a surge in sightings.

Sightings were up in 2020 compared to the previous year—with more than sixty-six hundred recorded during that period, according to National UFO Reporting Center data.⁴⁷

From the website How and Whys, which tracks this type of information, we get the following:

> William M. Tompkins, MUFON (Mutual UFO Network) Director: In July 2017, at a press conference, Tompkins made an unprecedented statement. He worked for the Douglas Aircraft Company alongside extraterrestrials (Nordic alien women). It had been four to seven years before NASA appeared, he claims in his book (mentioned above) that was published in 2015.⁴⁸

---

[47] Yaron Steinbuch, "Ex-CIA Director Believes UFOs Could Exist After Pal's Plane 'Paused'," *New York Post*, April 6, 2021, https://nypost.com/2021/04/06/former-cia-director-says-he-believes-ufos-could-exist-report/.

[48] Vicki Verma, "Earth Is Controlled by Warring Extraterrestrial Species,"

He met with two Andromedans who became his mentors and took him on board their space vehicles, including a large extraterrestrial mothership, where he was exposed to the teachings of the Andromedans over three months. The Andromedans gave him information about cosmic spirituality, life in the universe, and Earth's galactic history.

He claimed that his civilization, no matter where in the constellation it lives, has a single government and is spiritually forty-seven-hundred years more advanced than us, as well as five thousand years technologically. However, they maintain a balance of the spiritual and technological. In fact, they use technology to be able to evolve spiritually.

This is because there are different dimensional levels. We are currently in the third dimension. We still have not reached the fourth, much less the fifth. What differentiates the third from the fourth dimension is consciousness.

In the fourth dimension, there is a collective consciousness because everyone is telepathic and can read the minds of the rest. This leads to each person being authentic, transparent, and without ulterior motives. Upon entering this dimension, the being becomes clairvoyant since, in this state of consciousness, energy systems and

---

How and Whys, last updated February 27, 2024, https://howandwhys.com/william-tompkins-mufon-director-earth-is-controlled-by-warring-extraterrestrial-species/.

fields can be seen. (And here is the area of consciousness we are studying as it manifests itself in our reality, time-space, and other dimensions.)

However, there are still dualities, and the judgment systems are modified since the judiciary will be able to make decisions according to the energy fields. The beings of Andromeda have not lived in a society as manipulated as ours. They are telepathic and clairvoyant because they have studied all the sciences.

All souls know who they are; they all know about their past lives, and every time they incarnate, they are aware and know where they are going. In addition, they can see their evolution in the afterlife.

So, what is true? Between the lack of information, misinformation, and disinformation, the truth needs a friend. Am I to believe that aliens are hiding in plain sight and we as humans are a bunch of warring tribes, that heaven exists fourth-dimensionally, and that we are being genetically manipulated? Humans have ridden horses for the last twelve thousand years, and in the last one hundred, we went from riding horses to going to the moon.

How much of this is believable? That is the best explanation we have, and that is not enough. This needs to be studied, and formalizing the study will change the world. In terms of the study of consciousness, we must look at what we see when we use our third eye.

This goes to a very core area of consciousness: connecting individual consciousness with universal consciousness as a means of seeing. Telepathic ability is regularly encountered when dealing with aliens. They have a higher degree of extrasensory perception (ESP).

So, let us walk this path through the mind of the man who taught the US Army how to remote-view, Ingo Swann. In his book *Penetration*, he talks about the difference between Earthside ESP and spaceside ESP. When he talks about consciousness, he refers not just to individual consciousness or its connection to universal consciousness but also to how our collective consciousness is being manipulated.

Listening to Ingo's words is like visiting a pristine mountain stream before it has been polluted by travelers, campers, and con men. A fresh look at new data through the lens of Ingo's impressions yields strategies for understanding human consciousness and our place in the universe.

You cannot take part of Ingo Swann. By that, I mean you cannot accept him as the man who taught us how to remote-view and opened a means of data acquisition and transfer without also taking his description of aliens living right beside us and, it seems, managing their interactions with us. What I want you to keep in mind for the moment is that there are competing groups of aliens with different agendas. I would also argue that the US government is overmatched, as is humanity. But let us stay on topic.

The data on telepathy, which is how these aliens

communicate, based on exit interviews with abductees, gives the person with that ability access to information on a universal level. From the alien point of view, allowing human telepathy may be like giving a loaded gun to a child. In the next chapter, I will explain how we are beginning to navigate that space and how each of us, including you, my reader, can do it with the proper training.

We have a category in consciousness science calls "remote viewers," who have an ability to use psychic ability to see things in other places and times. They are, in a real sense, manipulating their energy.

Let me explain what it looks like to run a psychic mission with a group of remote viewers. The following is an actual example: An agency comes to the remote viewers with a specific task, in this case, NASA. They wanted Ingo Swann to look at Jupiter.

Swann worked for the Stanford Research Institute. In 1973 there was an experiment where Swann was asked to project his consciousness to the planet Jupiter. The reason Jupiter was chosen was because NASA had probes on the way there at that point. Swann made several observations that the probes soon confirmed: He saw a hydrogen mantle, high infrared readings, and the colors of the clouds. He also saw rings, which, at that point, no one had known about. He said they were smaller and closer to the planet than Saturn's and were made up of dust and tiny asteroids. This was not something we could see from Earth or even assume was there. The Pioneer probes in 1973 and 1974 did not see rings, but when Voyager 1 passed by Jupiter in 1979, it did see the rings as Swann described.[49]

Ingo also claimed that the moon is inhabited by a space-traveling, underground-living humanoid species that farms on the surface of the moon, in the Plato crater and Aristoteles crater. What I want you to envision is a twenty-story apartment building under each of those craters, with an inverted Coke bottle top where farming is done. Recent footage just released shows a UFO disappearing into a crater hole in the moon.[50]

I want to pause the dialogue just for a second. My editor wanted me to footnote this last statement. Rather than go back to my original notes, I just googled it. There are several videos of just this. I am putting the general link into the footnote because I want my reader to see the myriads of videos of UAPs coming and going. This is disclosure. There is a difference between disclosure and governmental admission of disclosure. They do not admit it because they don't know what to do about it. They do not admit it because they are making money off the technology. This book is about helping them help all of us.

---

[49] "Alien Bases on the Moon: The Amazing True Story of Ingo Swann," The Why Files, posted March 31, 2022, YouTube, https://www.youtube.com/watch?v=V8kT6J_uoic.

[50] YouTube, https://www.youtube.com/results?search_query=footage+of+a+UFO+disappearing+into+a+crater+hole+in+the+moon

Earth is isolated in a populated universe, and for us to join this galactic community, we must travel on the same 4D wavelengths aliens commonly operate on to protect us all. In my youth that attitude of those who control the little alien technology they have would be called penny wise and pound foolish. OK, the aside is over; back to the story.

In the underlying data, there are colonies of humanoids living on the moon, one of the moons of Saturn, and at the South Pole on Earth.

When Buzz Aldrin was on the moon, he could see these aliens. He switched to the medical channel and spoke to the chief medical officer. "They are here. They are parked on the side of the crater. They are watching us."[51]

Ingo's case is that as we approached the moon, we found extraterrestrials living there, and they said they did not want us there. Then they manipulated information so that we turned away from going to the moon and, in that manipulation, hid their existence from the general Earth population. Both America and Russia—at the same time, for no reason—stopped racing each other to the moon and jointly built Skylab, a more costly project for going to Mars. There was no reason for both adversaries to do this, except for Ingo's reason. The race to the

---

[51] "Buzz Aldrin Confirms UFO Sighting in Syfy's *Aliens on the Moon*," Entertainment Tonight, posted July 16, 2014, YouTube, https://www.youtube.com/watch?v=ZNkmhY_ju8o.

The Applications of the Science of Reincarnation

moon began as a competition between the Soviets and America in the 1960s, but by 1972, both countries had abandoned the moon as a goal; this in the face of the fact that we learned the moon had water, and you could grow crops in the moon soil we brought home.

Here, we get into the study of consciousness: how the aliens achieved their goal as seen by the man who was the one to teach us all how to remote-view.

Ingo states,

"It is far more likely that a concerted, and rather successful, attempt was undertaken with regard to TWO principal functions:

1. To increase rather than decrease space age confusions, so as better to promulgate and rule via disinformation packages.
2. To erect and reinforce a particular kind of planet-wide intellectual phase-locking that is data deficient with regard to the meaning not of Earthside affairs, but with regard to the meaning of Spaceside activities.[52]

"More simply put, if groups of Earthsider individuals can be brought, one way or another, into agreement

---

[52] Ingo Swann, *Penetration: Special Edition Updated: The Question of Extraterrestrial and Human Telepathy* (Swann-Ryder Productions, 2020) p139–140.

about the meaning of something, then their communal intellectual process will phase-lock with each other.

> Groupthink can then be formed with respect to this or that information package resulting in that intellectual phenomenon earlier referred to as mindsets."[53]

This is to say that Earthsiders do not think outside of Earthside local realities. This is further to suggest that the realities of Spacesiders might not fit into ANY recombination of Earthside information packages—and especially so IF Earthside intellectual phase-locking is deficient with regard to any Spaceside realities except those officially admitted to by science.[54]

Or, as we have seen, stage-managed by NASA.

Ingo adds, "Thus, more precisely defined, packaged information is meaning-managed."[55]

Yeah, that says it. That is consciousness, our consciousness, operating on itself.

So here is the consciousness witches' brew that is

---

[53] Ingo Swann, *Penetration: Special Edition Updated: The Question of Extraterrestrial and Human Telepathy* (Swann-Ryder Productions, 2020) p139

[54] Ingo Swann, *Penetration: Special Edition Updated: The Question of Extraterrestrial and Human Telepathy* (Swann-Ryder Productions, 2020) p141

[55] Ingo Swann, *Penetration: Special Edition Updated: The Question of Extraterrestrial and Human Telepathy* (Swann-Ryder Productions, 2020).

being served in this request: That belief in heaven, your heaven, is proven, as is your life after death; however, it is simply 4D consciousness, and it is replete with anything you see in three dimensions, including animals and aliens. It is a place of no lies; when communication is telepathic, you see the truth, so deception, prejudice, and such, as well as belief in a god, end because God is the collective mind. That 4D space is traversed by higher-dimension intelligence as well. There is a lot to study here, and consciousness science needs the funding.

Earthside Telepathy Versus Spaceside Telepathy

There is a difference between Earthside telepathy and spaceside telepathy. Each operates the same, the difference being in the case of the second, we are being encouraged not to use it by manipulation of the general consciousness of man. In short, that information is being societally repressed by being data deficient. But it is the undeniable alien presence in our skies that makes this area of study in such need of support.

If you want to study alien consciousness as it relates to human consciousness, begin with their better sense of telepathy, which we—as is demonstrated by the communication reported in abductions—show is our lesser ability.

We need to protect humanity, and we cannot do it with weapons. We, as a race, are third-eye blind, yet we have proven third-eye data acquisition beyond any reasonable doubt.

The point here to all concerned is that to have the technology transfer you need, you must be mature enough to use the technology. These races are thousands of years beyond us. We are suggesting ways to manage not just the technology transfer but also the cultural transfer to render humanity mature enough to handle the technology. With the speed of technology change, we can say aliens are in our workspace and are hiding in plain sight. Our understanding of consciousness must evolve at the same speed as the technology keeping both us and the aliens safe.

If we are going to package information and manage its meaning, teaching this to the next generation globally is critical. Funding this is not just about the money but also about validating the line of research. That the contribution has huge significance to social causes and economic opportunity, not just for your supporters but also for your opponents. Additionally, your supporters and opponents are now fused in a common goal/defense; we are all in this together, whether we like it or each other. Funny, isn't it? But there are headwinds. According to Ingo,

> "Telepathy is the most forbidden element of Earthside consciousness. Indeed, so forbidden that science would rather accept reincarnation,

the existence of the soul, and life after death—PROVIDED those situations DID NOT include any telepathic possibility.

Why this is the case is but a small tip of a gigantic iceberg."[56]

This is important because it denies an aspect of human consciousness that connects us. This denial is harmful to us individually and societally. The immense danger to this is it allows us to be controlled.

"Another amusing aspect is that the funding agencies did sponsor the secret developmental work in remote viewing—somewhat on the grounds that it penetrates things, not minds.

This is to say that remote viewing pertains to penetration of 'physicals,' not to penetration of 'mentals.'

In any event, the principal reason why ALL formats of Psi research are marginalized, treated to energetic diminishment, or suppressed altogether is that those formats do include potentials too near the hated and unwanted telepathic faculties.

So, the whole barn of psychic research must be burnt down as quickly as possible, making sure that the telepathic horses don't escape.

There is one notable exception to this, and one

---

[56] Ingo Swann, *Penetration: Special Edition Updated: The Question of Extraterrestrial and Human Telepathy* (Swann-Ryder Productions, 2020), 144.

utilized for creative cover-up purposes. This exception involves the discovery of approaches to telepathy most noted either for the fact that they DO NOT work, or because they serve to disorient and defeat approaches that MIGHT work.

Thus, the concept that telepathy is a mind-to-mind thing involving a sender and a receiver has been given extraordinary publicity—and has in fact become the principal Earth-side cultural model for it."[57]

At least two observations can be made relevant to the above. First, one might consider that the Earthside retreat from Psi is something akin to protesting too much. Second, if I were an extraterrestrial with highly developed Psi skills (and which might have led in the first place to the evolution of superior technology), I wouldn't particularly want Earthsiders to develop Psi faculties. And if telepathy was an element in, say, consciousness universal, I would soon figure out how to telepathically impregnate Earthside human consciousness with intellectual phase-locking that was detrimental to positive telepathic-plus development. The reason might be very obvious. After all, what extraterrestrial would want Earthside telepaths penetrating spaceside affairs, especially, perhaps on

---

[57] Ingo Swann, *Penetration: Special Edition Updated: The Question of Extraterrestrial and Human Telepathy* (Swann-Ryder Productions, 2020), 146.

the moon so near to them? Thus, in this, at least, spacesiders and Earthsiders might have something in common—the Telepathy War, won hands down so far by the Spacesiders.

If the existence of groupthink and intellectual phase-locking is accepted, then the only remaining problem, or opportunity, is what information packages are to be inserted into them and thereafter managed for one end or another. However, in the light of the above, the existence of group-minds cannot be escaped.

There is another reason why disclosure has not occurred.
We can accept the fact the aliens wish to remain anonymous while even helping us with leaps in technology. However, our own people put the cork in the bottle to prevent and retard disclosure.

President Harry Truman put into place an entitlement program that was clandestine and does not report to Congress by its very charter. I am not going to dwell on this point because frankly, I feel it would not be safe for me. The following is a public document. If Truman did indeed authorize Operation Majestic 12, it still operates today. It may have morphed in structure or name, but whatever is operating today, retarding public knowledge, has its roots in this act.

Before you, my reader, conclude that I am in some way indicting these people, I would tell you to look at what they had on their plate. Their mission was to protect humanity. The actions taken were in our interests at the time, and the politicians' inability to deal with what they saw as a threat that they would have to admit was beyond their ability to respond to.

On December 12, 1960, FBI special agents interviewed former president Harry Truman, at the request of President-Elect Kennedy, and Truman was "cordial until purpose of interview explained, whereupon his manner became brusque." Was he asked the embarrassing question, "Did you authorize Operation Majestic 12?"

The following document is from the FBI Vault archive:

Record of Interview of former president Harry S. Truman

From the interview body, it appears that the incoming administration of President-Elect John F. Kennedy requested the Special Agent in Charge of the Kansas City FBI Office to interview former president Harry S. Truman over a "special inquiry matter." Truman was clearly upset at the line of inquiry and eventually terminated it, after his opinions were asked about certain people that were to hold high-level positions in his government. As the Director of Central Intelligence is a political appointee, I believe one of those people was the incumbent DCI at the time, Allen W. Dulles. Truman appointed Dulles as Deputy Director of

The Applications of the Science of Reincarnation

Central Intelligence in 1951, and tells the FBI special agent he is "all right." The other person that Truman claims he "never heard of" may have been CIA Director of Counterintelligence James Angleton, who was appointed to that position after Truman had left office. Kennedy may have heard bad things about these two gentlemen through his friend James Forrestal, whose defenestration in 1949 haunted Kennedy for the rest of his life.

Interestingly, another reference to Truman being interviewed by FBI special agents regarding "special inquiry matters" comes from the Majestic Documents via Source S-1, whom I believe to be CIA CI Special Investigations Group Director Newton "Scotty" Miller. In this interview, Truman is asked directly, "Did you authorize Operation Majestic 12?" Truman is initially coy in his response until he is shown the documents with his signature on them—supplied to the FBI by the CIA. He then proceeds to expand on the reasons why it was created and how Congress had no idea of the monster they had all created. Miller supplied his essay "UFOs, CIA and Congress" to researcher Timothy Cooper in 1999, and the discrepancy in the years may have been due to a lapse of his memory.[58]

---

[58] "Unidentified Flying Objects, the CIA, and Congress," Majesticdocuments. com The original FBI record of interview file can be found on page 54 of

Not Disclosure but Context

As I stated in the post, Miller might be getting the December 1960 FBI interview mixed up with the December 1963 *Washington Post* article that Truman wrote calling for the CIA to be "limited to intelligence gathering." The fact Truman had it published precisely one month after Kennedy's assassination caused Dulles and Angleton to fly to Independence, Missouri, the next day to get Truman to rescind the article. Truman calmly and politely told both men to "f*ck off."[59,60]

Homer, in this chapter we learned that several different kinds of aliens are here; the government knows, and they are not telling you. That is what data deficient means, Homer; it means they do not want you to know.

What don't they want you to know? That there is a shadow government inside the United States dealing directly with aliens.

---

this FBI Vault Records: The Vault-Harry S. Truman part 02of 03, https://majesticdocuments.com/pdf/ufos-cia-congress-s1-00.pdf.

[59] Harry S. Truman, "Limit CIA Role to Intelligence," *The Washington Post*, December 22, 1963, https://archive.org/details/LimitCIARoleToIntelligenceByHarrySTruman/mode/2up.

[60] "On 12 December 1960, FBI Special Agents interviewed former President Harry Truman at the request of President Elect Kennedy and was 'cordial until purpose of interview explained, whereupon his manner became brusque.' Was he asked the embarrassing question 'did you authorize Operation Majestic 12?'," Reddit, June 2024, https://www.reddit.com/r/UFOB/comments/1dppm9o/on_12_december_1960_fbi_special_agents/?share_id=xmprkhXpxVRcshaqFGjzY&utm_content=1&utm_medium=ios_app&utm_name=ioscss&utm_source=share&utm_term=10.

Chapter 7

Disclosure: It Has Happened

Is there such a thing as the common mind? If there is, how would it work? If you are electromagnetic then you can be a particle or a wave or both simultaneously. If you are "on the same wavelength," you are connected to the common mind as we explained in NDEs, children who remember prior lives, past-life regression, but also remote viewing and telepathy. But can the common mind be influenced collectively? If it can, then there lies the danger of all of us being manipulated. Ingo Swann, in his book *Penetration*, asks the same question.

Is Consciousness Individual or Universal?

Ingo says,

> My own research into this area revealed that mass consciousness or mob consciousness research came to an abrupt end in about 1933–1935. This is to say that it came to an end as far as public access to it is considered.

It ended because of a set of discovered conclusions. Among them, that mob consciousness responded collectively NOT to rational intellectual perspectives, but to some kind of emotional empathy that was somehow subconsciously TRANSMITTED. This, however, could not be explained unless the concept of telepathy was brought into consideration.[61]

I agree with Ingo about the study of mass consciousness ending abruptly. Let us meet the man and the reason why. In his world the terms "public relations" and "propaganda" were synonymous. At this point in the narrative, I want to introduce Edward Bernays (1891–1995), an American who was considered a pioneer in the field of public relations and propaganda and was referred to in his obituary as "the father of public relations." He was seminal in the fields of public relations and propaganda. The following is from Wikipedia.[62]

Public Relations (1945) outlines the science of managing information released to the public by an organization, in a manner most advantageous to the organization. He does this by first providing

---

[61] Ingo Swann, *Penetration, The Question of Extraterrestrial and Human Telepathy* (Swann-Ryder Productions, 2020), page 165.

[62] "Edward Bernays," Wikipedia, https://en.wikipedia.org/wiki/Edward_Bernays.

an overview of the history of public relations, and then provides insight into its application...

But instead of a mind, universal literacy has given [the common man] a rubber stamp, a rubber stamp inked with advertising slogans, with editorials, with published scientific data, with the trivialities of tabloids and the profundities of history, but quite innocent of original thought. Each man's rubber stamp is the twin of millions of others, so that when these millions are exposed to the same stimuli, all receive identical imprints...

Bernays touted the idea that the "masses" are driven by factors outside their conscious understanding, and therefore that their minds can and should be manipulated by the capable few. "Intelligent men must realize that propaganda is the modern instrument by which they can fight for productive ends and help to bring order out of chaos...

Propaganda was portrayed as the only alternative to chaos...

One way Bernays reconciled manipulation with liberalism was his claim that the human masses would inevitably succumb to manipulation—and therefore the good propagandists could compete with the evil, without incurring any marginal moral cost. In his view, "the minority which uses this power is increasingly intelligent and works more and more on behalf of ideas that are socially constructive."

Unless they are not—or used by smarter societies to control humans. Divide and conquer, or divide and control; it is still manipulation.

That is why everyone in America who claimed they saw a UFO was considered a "kook" or "nuts." That was in the 1950s. That was the information package inserted into the groupthink at the time.

This has been refined today by firms such as Facebook, Reddit, and Fox News. They stratify society and make it dumber, while computer AI gets smarter.

The best way to break this roadblock is to explain and study consciousness on a 4D level. The change begins with breaking the phase-lock to open 4D studies. How are the best alien societies modeled? How can we effect positive global change? How do the alien societies we see and now acknowledge flying by us administer governments? What are their rules? Targeting and then teaching remote viewers how to work in unison is by nature a collaborative effort and larger scale because of the number of people involved.

Using 4D communication, remote viewers can bring back credible information. Now group them to increase bandwidth and begin to query global management questions and repair of our planet's resources. These are programs that need to be run.

How did alien societies that we see that travel the stars get past that point of self-harm? The means of communication are open, not with radio frequencies but target group consciousness.

Pause for a moment and think of how this narrative changes the narratives of religion, gender, politics, and self-governance. We are all victims of this new reality. It favors no one, but if we are all threatened, we are all in this together.

Given the new emerging conditions, the following is logically deduced. It is opposed by the phase-lock groupthink.

Bernays developed ad campaigns for businesses, cigarettes, and so forth. He was Freud's nephew, and while Freud is arguably more famous, Bernays is far more important and directly relates to the topic of aliens.

You see, it was not just a product he was selling; he defined mass marketing and control of information for governments. What you tell a group of people and how you tell it can send countries to war. In 1954 a democratically elected government in Guatemala was overthrown by the CIA using a few radio stations. The radio stations controlled by the CIA began broadcasting that an independent army was fighting to liberate Nicaragua from the government. The government could not find where the fighting was occurring because there was no fighting.

Ultimately the government fell because people across Nicaragua believed the story.[63]

Now let us look at our situation on Earth. Does anybody think that aliens, which are demonstrably here, cannot insert false information into our media stream and control our beliefs? We had proof, unequivocal proof, of aliens in 1947 in the United States because they crashed here. We had their bodies and their ships and began negotiations, which we will detail in the coming pages.

By all measures, aliens have been here for at least 12,500 years to put a satellite we call the Black Knight satellite in polar orbit around Earth. Again, by all measures, there is proof of collective consciousness, which we are seeing in their ability to communicate telepathically and collectively. It is an ability that humans have that remains underdeveloped. We see this ability in psychics and remote viewers, and it manifests itself in near-death experiences (NDEs), past-life regression, and children who remember prior lives (CWRPL). Tying these apparent anomalies together redefines our view of the universe and ourselves and puts us in some spaces where we are clearly unwelcome.

The threats we define here and are addressing are not apparently the aliens, who all seem to have their own agenda. After all, if they wanted us gone, they have had at

---

[63] "1954 Guatemalan coup d'etat," Wikipedia, https://en.wikipedia.org/wiki/1954_Guatemalan_coup_d%27%C3%A9tat.

least 12,500 years to do it. So, make no mistake about the information we are to traverse here. This book is about their intent and ours. How do we collectively respond to this?

Funding the study of our consciousness should be our first defense, and as we begin to navigate the exopolitics—the politics of alien societies interacting with Earth—we need to build new models of social understanding, so we remove blocks that prevent us from uniting us psychologically.

If our very energy can interlock, then blocks of interference such as race, color, gender, and religion need to be modulated and mitigated without a global conflagration during which we destroy ourselves.

Again, by any measure, there are a variety of reasons why humans have been allowed to develop, and it seems there are alien groups who have a very strong interest in our DNA and other assorted genetic material. Again, by the measures we have, the Grays, an alien race, have abducted more than a million people, and the Eisenhower administration had an agreement with them for the Grays to give them lists of names.

Now here is the problem for you, my reader. If you look at the underlying proof, you will see this is true, and in the coming few years, NASA will admit we have alien contact. When they do, the information will be gradually revealed to you. This book is about what we, the people, can do to address the failure of our governments and, at the same time, help those same governments with a problem we all face.

So, we will try to organize a hypothetical group of billionaires to work as a team to address these issues.

This entire structure and course of study must be laid out in a single compendium, studied, and taught. So yes, good scientists get money, but in doing so, science is advanced, humanity is strengthened, and acknowledgment is the only requirement; your money does that.

I need a 4D workshop that draws on 7.5 billion minds in unison, working cohesively on one project, and I might get somewhere. Cohesively focusing the mind with a group increases bandwidth. With 7.5 billion minds, we do not know what can be achieved until we try.

All aspects of human sexuality and experience are accepted and equal from right to left, from up to down, and from black to white. What needs to be done so we, as the human species, can join the galactic community?

Ingo says,

> "As it is, telepathy cannot exist, much less be explained, IF the parameters of consciousness are limited to the mental equipment of the biological individual.
>
> Since information is "exchanged" or "acquired" between human individuals in the absence of any objective methods to do so, and in that the

information so exchanged results in mental perception of it, it is obvious that a format of consciousness exists that is independent of each biological human unit."[64]

"We might also have to wonder if their telepathy is a developed version of a telepathic "language" that is universal within universal consciousness."[65]

"A larger, much larger question eventually loomed into view: Why do mass-consciousness humans, as it were, mass-consciously almost "conspire" to avoid certain issues, and consistently so? My investigations into this matter have revealed that four general areas of societal avoidance have existed for quite some time:

1. ***Sexuality and eroticism.***
2. ***Human psychic phenomena.***
3. ***General societal love.***
4. ***UFOs and Extraterrestrials.***[66] [emphasis mine]

Within human psychic phenomena, you encounter aliens, and sexuality and eroticism morph into general

---

[64] Ingo Swann, *Penetration, The Question of Extraterrestrial and Human Telepathy* (Swann-Ryder Productions, 2020), Page 167
[65] Ingo Swann, *Penetration, The Question of Extraterrestrial and Human Telepathy* (Swann-Ryder Productions, 2020), Page 168
[66] Ingo Swann, *Penetration, The Question of Extraterrestrial and Human Telepathy* (Swann-Ryder Productions, 2020), Page 225

The Applications of the Science of Reincarnation

societal love as the fact that we travel lifetimes as individuals of both sexes. Human psychic phenomena are part and parcel of human consciousness, and we must be bold enough to acknowledge them and study them in an open and robust manner if we are to save this planet.

Ingo adds, "It is with good reason I believe, hitherto almost unimagined, that all four of these areas are at least linked with regard to an Extraterrestrial abductee context which is positively awash with sexuality overtones, while the psychic nature of UFO abductee experiences is visible beyond argument."

Information packages that have been kept apart should be fused in our examination of consciousness, both individual and group mind.

Ingo says, "A large and vivid vacuum of information exists regarding the phenomena of group mind and subliminal group consciousness management that might be invasively influenced by various means such as forms of super-telepathy as yet unacknowledged as existing."

So, what can we do?

The phase-lock must be broken. The aliens are not just here; they are among us, and we carry their DNA.

We must manage global contact with alien civilizations outside the government.

No one in any government wants to be the one to

officially recognize aliens, and no one person can do this alone.

One method of contact is crop circles, which is open to everyone.

Laurance Rockefeller has invested in the study of crop circles,[67] and clearly, many are of alien origin. We know this because of how the stalks are broken, compared to the fakes. At its core, what we are seeing is both an information package and a disinformation package arriving at the same time, one covering the other. Rockefeller's research needs to be brought to the consciousness center. The alien crop circles, in many cases, are fractals, which carry information inside the information so the growing resources of the consciousness center can be brought to bear on this valid method of alien communication.

We now arrive at the most important video connection in this paper. It is only six minutes long. It explains what is happening with crop circles, and since we sent a message into the universe and, twenty-seven years later, got a response, what do we say now? Apparently, there are those who, by discrediting the validity of this means of communication, would have us remain unaware of our situation and reality.

---

[67] Constance Holden, "Rockefeller Finances Crop Circle Survey," Science, May 21, 1999, https://www.science.org/content/article/rockefeller-finances-crop-circle-survey.

Crop circles are misunderstood, and information that should be public and primary is obfuscated and withheld from a variety of actors. In a very short order, we are going to accurately explain crop circles, expose the bad actors and intentions, and arrive at a solution/conclusion.

Mr. Mark Cuban, please follow this thread to its conclusion, as it will help me explain why you should sit on the Applied Technology Committee, chair of pharmacology. This is of the utmost importance. The story begins on November 16, 1974, when Frank Drake, known for the Drake equation, which estimates the probability of life beyond Earth, and noted other scientists, including Carl Sagan, broadcast a message from the Arecibo radio telescope.

This binary-coded message was sent into space from the Arecibo radio telescope in Puerto Rico. Interestingly, and for no apparent reason, this message was pointed at the Grays' home planet. The videos below will explain the process. The point here is that twenty-seven years later, outside the Chilbolton radio telescope in England, there appeared new crop circles. One, however, was a direct correlation to Sagan's message.

This is a picture of the message in binary code. On the left is the message Sagan and the other scientists sent years before. On the right is the message that appeared as a crop circle in front of Britain's largest radio telescope, the Chilbolton radio telescope.

This page intentionally left blank

ARECIBO Nov. 16, 1974 Trans.

0	0	0	1	1	1	1	0	0	0	**NUMBERS**
1	0	0	1	1	0	0	1	1	0	1-10
10	11	10	1	0	1	0	1	0	1	(right to left)

11000	**ATOMIC NUMBERS FOR LIFE**
10110	(from right to left) Hydrogen 1;
10110	Carbon 6; Nitrogen 7; Oxygen 8
10101	Phosphorus 15

Right to le

00011	00011	00111	00011	Deoxyribo
00001	01100	00000	00001	ADENINE
01011	00011	00110	01011	THYMINE
01000			01000	Deoxyribo
00000			00000	Phosphate
10000			10000	Phosphate

00011	00111	00011	00011	Deoxyribo
00001	00000	00100	00001	CYTOSIN
01011	01110	01100	01011	GUANINE
				Deoxyribo
01000		11	01000	Phosphate
00000		11	00000	Phosphate
10000		11	10000	

NUMBER OF BASE PAIRS IN HUMAN DNA 4,294,441,822

11
11
10
11
11
10
11

MOLECULAR STRUCTURE OF HUMAN DNA AN NUCLEOTIDES;
C5OH7; C5H4N5
C5H5N2O2; O4P
C4H4N3O;
C5H4N5O

DOUBLE DNA HELIX

VISUAL REP. OF A HUMAN BODY

10
11
01
11
11
10

1101
1111
1110
1101
1111

— 1110
AVG. HEIGHT OF HUMAN 5' 9.5"

HUMAN POPULATION IN 1974 4,292,853,750

VISUAL REP. OF SOLAR SYSTEM
(right to left) Sun; Mercury; Venus;
Earth; Mars; Jupiter; Saturn; Uranus;
Neptune; Pluto

VISUAL REPRESENTATION OF ARECIBO TELESCOPE IN PUERTO RICO (1974)

ACTUAL DIAMETER OF ARECIBO TELESCOPE 1004.52 Feet

101001 = 2430 | 12.6 cm * 2430
011111   30,618cm = 1004.52'

```
1 ——————X—————— 23
```

Chilbolton August, 2001 Template

```
0  0   0 1 1 1 1 0 0 0     NUMBERS
1  0   0 1 1 0 0 1 1 0     1-10
10 11 10 1 0 1 0 1 0 1    (right to left)
```

```
111000   ATOMIC NUMBERS FOR LIFE
110110   (from right to left) Hydrogen 1;
110110   Carbon 6; Nitrogen 7; Oxygen 8;
100101   Silicon 14; Phosphorus 15
```

Right to left

00011	00011	00111	00011	Deoxyribose;
00001	01100	00000	00001	ADENINE;
01011	00011	00110	01011	THYMINE;
01000			01000	Deoxyribose;
00000			00000	Si Oxygen 4;
10000			10000	Si Oxygen 4

00011	00111	00011	00011	Deoxyribose;
00001	00000	00100	00001	CYTOSINE;
01011	01110	01100	01011	GUANINE;
				Deoxyribose;
01000		11	01000	Si Oxygen 4;
00000		11	00000	Si Oxygen 4
10000		11	10000	

```
                         11
NUMBER OF                11
BASE PAIRS               11    MOLECULAR
IN ALIEN DNA             10    STRUCTURE OF
4,294,966,110            11    ALIEN DNA AND
                         11    NUCLEOTIDES;
ALIEN DNA                10    C5OH7; C5H4N5;
VISUAL REP.              11    C5H5N2O2; SiO4;
                         10    C4H4N3O;
VISUAL REP.              11    C5H4N5O
OF AN ALIEN              11
BODY                     11
                         11
                         11                111110
                         10                111101
                     ———ALIEN               101011
             ——1000     POPULATION         011111
         AVG. HEIGHT    IN 2001            110111
         OF ALIEN 3' 4" 12,742,213,502       1011
```

VISUAL REP. OF SOLAR SYSTEM
(right to left) Sun; Mercury; Venus;
Earth; Mars; Jupiter Moons; Saturn;
Uranus; Neptune; Pluto

VISUAL REPRESENTATION OF
ALIEN COMMUNICATION DEVICE
WHICH APPEARED IN AUGUST 2000
NEXT TO CHILBOLTON TELESCOPE

ACTUAL DIAMETER / WIDTH OF ALIEN
COMMUNICATION DEVICE 2789.52 Feet

```
0010110 = 6748 | 12.6 cm * 6748 =
0011101   85,024.58cm = 2789.52'
```

From Pinterest[68]

This is disclosure, not from our government but from this group of aliens specifically. It is not a one-off, but a channel opened that the government cannot conceal because of the size of the crop circles, but the circles themselves can last more than a year.

The first line of Sagan's message was our decimal system. The first line of their message was the decimal system, confirming we were speaking in the same mathematical language.

In Sagan's message, the next line was our dominant element, which was carbon. In their second line, it was the chemical code for silicon, their dominant element.

In the subsequent lines, which you will see in the video, they are four feet tall, have large heads, and live on the second, third, and fourth planets of their solar system. While Sagan told them there were eight billion humans, there are twelve billion of these aliens, according to their own statement.

Finally, the last line was the picture of our radio telescope, and in their response was a picture of their radio telescope, which incidentally corresponds to another crop circle of just their equipment.

I have two six-minute videos for you to watch that explain crop circles. The first explains what is going on,

---

[68] https://www.pinterest.com/pin/610800768248364148/

Disclosure: It Has Happened

whom we are communicating with, and where they are from. The second is what a real crop circle, newly made, looks like.[69,70] The third video goes into more detail and is twelve minutes long.[71] On the bottom of each image is a picture of radio telescopes: on the left, what the Earth's radio telescope looks like, and on the right, what the aliens radio telescope looked like. In the next line up, the black square is the sun. On Earth's side the third planet is pictured, representing the planet on which we live, and the larger squares after Earth represent the gas giants, indicating that we have nine planets in our solar system.

In the right planets three and four are raised, indicating they live on two planets in their solar system. The one image after that we do not understand is perhaps a satellite where people live. They have eight planets in their solar system, and going back to their sun, it is smaller than ours. I am not going to depict this entire communication, but we said there are 8 billion of us, and they said there are 12.7 billion of them.

I want to pause for a second and plant a seed. They have space travel, more headcount than us, and could

---

[69] "Alien Civilization makes contact! Courtesy of Terje Toftenes & UFO TV," Andy Bell, posted on June 10, 2012, YouTube, https://www.youtube.com/watch?v=G-JU8laFmYw.

[70] "Crop Circle," July 27, 2010, Media Tour, posted September 22, 2019, YouTube, https://www.youtube.com/watch?v=jC-OpdxmS7I.

[71] "Arecibo Message (Detailed Explanation)," Science World, posted January 27, 2021, YouTube, https://www.youtube.com/watch?v=HYuCr6Er914.

have taken over our planet thousands of years ago by stamping out primitive man. Why didn't they? Perhaps humanity itself is a product that is maturing. We are more of a danger to us than they are and will not help unless asked. This book is about a proposal not just to billionaires but also to the aliens themselves.

You see, there is a direct, open channel to these aliens, and they are communicating solutions to Earth. Mark, can you imagine if there was an open program to ask? The next chapter explains why.

In the videos look at the detail of the information contained in a real crop circle, and the intricate way the stalks are tied; this is all information that needs to be decoded by our scientists.

Going through other messages that have been sent this way, you find this alien species referring to another alien species from another narrative we are presenting here. Their comments flawlessly blend with the other narrative, and both are discredited by our authorities.

If you google crop circles, the first thing you see is the vast amount of information discrediting what I just showed you. When I say it is a bad narrative, I think the discrediting group could have done better than two guys with boards. Look at the detail of the real crop circles and look at the cover-up. Can two men with boards do this?

The lie is so transparent and broken that this information must be addressed cohesively, collectively, intelligently, and responsibly, and that is just not happening.

Is this common knowledge? NASA cannot find the aliens; here are twelve billion of them, and I know where they are. This is happening in England, outside NASA's jurisdiction, so we cannot blame them. But if you are looking for aliens, I would start with the ones putting up the billboards in our wheat fields. Oh, another thing, can we open the real information without the disinformation overlay so we can bring more resources to bear on decoding the layered math in how the wheat stalks are being laid down? Bring that information to the math departments within the consortium. Please. Thank you. If my readers want proof, they can watch the videos I posted above, read the science journals, or read about Sagan. This is a WTF moment.

NASA has asked that we listen to space. The link for this is in chapter 12, "Theory Development." However, remote viewing is not encouraged outside the control of the CIA or the intelligence services.

Remote hearing is an area of consciousness studies that may solve this problem for NASA. How does the science community and the NGO community bring resources to bear for an organization that has lied to us for seventy-five years? I understand there will be political fallout from the book being requested from the scientists, *The Standard Model of Consciousness*. That very idea reorders the political and social landscape. But your support of this proposal brings real funded organizations to the reality that we must face whether we can acknowledge it or not. Do not send us on a snipe hunt when you want us to bring back pheasants. You are asking for help. We want to help. We all need to be working on the same page of the playbook. Please, NASA, support this effort, and it will bring you all sorts of resources.

Here are the questions we should and must address:

Have we dissected the images of the alien equipment? Since the channel is open, can we ask questions or make the transmission of the conversation faster? Who is studying the images they send? Who is orchestrating the cover-up? Oh, there are so many more questions.

Well, it is Laurance Rockefeller funding the study, but shouldn't this be done with more people and more resources? Not if you are going to keep it a secret, or try to, which, by its design, limits the resources being brought to bear on a problem.

At this point this book is not even an exposé; it is a real question posed to the secret keepers: At this point, why? You need more resources; the two thousand billionaires represent half the world's wealth. This is new markets and new technology access, and aliens want the use of Earth's rich DNA pool, among other resources.

By opening this information to the public, common consciousness becomes more cohesive. To manipulate an entire species, control of the information becomes critical, and loss of control is threatening to the species attempting that control. Sites such as Reddit become a data-mining operation in terms of extraterrestrial and transdimensional contact.

Integrating a global response would seem a strategy that would benefit humanity, support our intelligence services, reduce hostility among us, and make us stronger.

That means scientists themselves need a collective voice so they cannot be "picked off" when presenting ideas outside the current paradigm of thought. That is why this plan, for all the blemishes that critics will find or attribute to it, needs support from the consciousness science community, which needs the support and freedom to do what the Unidentified Aerial Phenomena Task Force (UAPTF) is asking us to do—find new ways to address our problems. It starts with the government saying this is accurate information and owning the encounters until it hits their secrecy boundaries. Critical thinkers would understand full disclosure is not necessary, but disclosure is mandatory and, with it, an explanation of consciousness, so this new information comes with public reassurance. Here is the press release from the government; it is fictional, this is what I would want them to say.

> "Yes, we have had contact for the last seventy-five years; nothing here on Earth has changed, but technology has gotten better, and maybe we can stop killing each other; there are other species thousands of years beyond us, and they do not care about us. They have been here for twelve thousand years, and we might be able to learn something from them."

I do not want to put words in NASA's mouth, but if UAPTF is looking for new solutions, just give me some time on a radio telescope, and I will ask my galactic neighbors. I will also find out more about the other alien species.

At this point the lie that NASA cannot find aliens is so brittle and broken that coming clean will not be so big a deal. And doing it through consciousness science will allay fears and change social and political behavior.

This is unequivocally human-alien contact outside the auspices of NASA or any governmental agency. It is aliens communicating with humans and literally putting a billboard in front of the radio telescope. It is an open channel or conduit to communicate with an alien species. But their communication to us references another alien race that they say is misleading us. They are acknowledging that there has already been human-alien contact of a race other than them, and this drives home the need for exopolitical courses taught at our universities.

The disinformation campaign would have you believe two men with boards can make a crop circle, and they have. But they cannot make one like this overnight, and that is the point. This is real. Some of these crop circles are Julian sets, which are math fractals.[72] The information contained in them needs to be processed as a team, and by watching the video of the close-up of the crop

---

[72] "Julia Set," Wikipedia, https://en.wikipedia.org/wiki/Julia_set.

circle, you can understand just how much information is being conveyed. Please study this openly. Now that the line of communication has been established, what are you going to say next? Do you understand what they have said to you?

Scientists, do you want funds for studies? Publicize this and ask the public what should be said. Now as a strategy, you are normalizing the conversation and bringing humanity into an intergalactic mindset. You are also opening more lines to funding by more clearly articulating the problem. At its core, all this we are proposing is just other ways to study not just human consciousness but also all consciousness.

Now google crop circles, and what comes up is the misinformation overlay to discredit real communication. They are responding quite clearly to a message sent into the cosmos in 1974, written by Carl Sagan, among other scientists, including ones who worked at NASA.

So why the crop circle cover-up? Is the disinformation campaign diminishing our own ability to defend against extraterrestrial threats or to keep us separated?

What is collective human consciousness? What does it look like? How does it operate?

Here is an answer: Some of us know this is true; others of us know it is true and do not want others to know. But if we all knew, then we would have a collective response rather than a segmented response by a special few. Institutionalizing this teaching is a cognitive change.

That change of consciousness has geopolitical, exopolitical, and social ramifications. It is funny, really. Households will be asking, do you really want your daughter to marry a Martian? Or a Reptilian, for that matter? But normalizing that kind of question normalizes the human response to aliens as just being other humanoids, which they are.

Let us look at Sagan's message in the video. In binary code, which is explained, he stated that humans have DNA, the size of the average human, our location on our planet and solar system, and a diagram of the radio telescope we used to project this binary code into space. Twenty-seven years later, a crop circle appears that is the replica of Sagan's message. This alien race locates their planet from their sun; they live on the second, third, and fourth planets of their solar system; they list, like Sagan did, their total population; and lastly, and importantly for our discussion, they diagram their radio telescope.

Now we have a clearly defined line of communication with an alien species outside NASA. What do we do with it? My first question would be about expanding knowledge of their equipment to make the channel wider, but that is just me. What is our collective response? What are the lost opportunity costs? The messages, if you look at them and can filter out the two clowns with the boards who are providing the disinformation campaign, refer to another alien species here on Earth. If this is game on, then where is my team? It is segmented, creating lies and mistruths to hide it from the general population, thereby preventing a common consciousness response. If you do not think this is about funding the study of human consciousness, you are dead wrong. But if we are to do this honestly, then this is what we must face and acknowledge: the greatest threat to us may be ourselves.

This can all be changed by our scientists collectively acknowledging this reality and the billionaires independently funding the study of consciousness. While some intelligence service secrets will fall, the intelligence services will benefit far more from this change than any damage it may do in the short term. Scientists can explain this to them. The coalescence of the information—that we have one group of aliens here and another there—is not unprecedented in human history. Look to the British exploration societies of the late twentieth century that gave us great Hollywood exploration movies.

All the science here hangs together, and aliens are literally putting up roadside signs that are huge, intricate, and filled with information. Watch the video inside a fresh crop circle. Are we analyzing the way the crops are tied? Are there patterns? Are we looking for layered information? Are we addressing this collectively, cohesively, and intelligently?

By institutionalizing this teaching, it supports our intelligence services and our defense posture. It reduces friction between us and opening new intergalactic markets, which are closer than many think.

How?

Psychic energy in the brain connects with the psychic energy of the field for information transfer and access.

I cannot tell you there are labs that are decoding the human genome to find the gene responsible for enhanced remote-viewing perception. Changing and enhancing that DNA creates a new class of humans. But I should be able to. If I wanted to, maybe I could.

That DNA can be enhanced by AI to see deeper into the universe and our reality.

Human DNA is being manipulated with clustered regularly interspaced short palindromic repeats. If a lab has isolated the DNA marker for psychic ability, will the next generation of humans be able to have enhanced 4D sight?

The DNA identity of the psychic gene, its planned enhancement, and its need for funding are fundamental to consciousness research. This should be funded by more than dark money to establish human sovereignty over human genetics.

How much would that be worth if Bio Data Transfer Technologies (just an example, not a real company) became a company with the cutting edge of the world's leading scientists at the consciousness consortium and a freer information flow with the intelligence services and the Grays? Would it merit Venture Capital interest? But all the movement here needs to be integrated. There are exopolitical issues that need to be addressed with the type of proposal I am asking you to be involved with, the first of which is presenting this plan to them as it is. When you have the opportunity to read this, so will they.

This plan is asking for large amounts of money for the best consciousness scientists in the world. Why would the billionaires of the world invest in fundamental research they do not understand? The scientists cannot even make any guarantees about ROI.

Self-protection. From what?

I am going to explain where we were twenty years ago. You already know of the dangers of dealing with transdimensional beings. The same can be said of extraterrestrials.

The following is a short synopsis of the Reptilians:

> The conclusions Brown draws from his remote viewing the Reptilians are as follows: (1) The Reptilian ET species exists; (2) it is currently involved in a significant military confrontation with another group, possibly the Grays; (3) Earth and humans have been indirectly caught up in this military confrontation; (4) the reptilians have at least one underground facility on Earth in which both they and a Reptilian/human hybrid species work together; and (5) the reptilians have some future plan for earth and humans, although we do not yet know any specifics for this plan.[73]

---

[73] Courtney Brown, *Cosmic Explorers, Scientific Remote Viewing, Extraterrestrials, and a Message for Mankind* (Signet Books, NYC, NY 2000), 193.

This was published in 2000, around the same time Ingo Swann published *Penetration*. What is going on now? Even if I knew, I would not say, but something should be done about it. This plan that I am asking you to lead is that response.

You start with the billionaires I have listed here and send them this book, and as TIFACS evolves, the scientists will write their needs in a cohesive format.

Do you think transdimensionals are dangerous? So are the extraterrestrials.

Why is this important?

Let us look at three possible outcomes for humanity as outlined by the Farsight Institute, a remote-viewing organization, as viewed from the point of view of the aliens we are discussing.

The Reptilian agenda: According to Courtney Brown, "It appears that the agenda of the Reptilian Extraterrestrials is to use the genetic stock of humanity to create a new race that is partially Reptilian."[74]

This future for humanity has harsh features because once the new race is created, there is little use for the gene bank, which means humans. There is no future for humanity in the Reptilian agenda.

The Gray agenda: There are similarities in intent with the Reptilians; they are also using our gene bank, but

---

[74] Courtney Brown, *Cosmic Explorers, Scientific Remote Viewing, Extraterrestrials, and a Message for Mankind* (Penguin Putnam, NYC, NY 2000), 301.

there seems to be no desire to hurt or subjugate humans; they would prefer our free-given friendship.

The Federation agenda: The best way to explain the federation view of humans is a laissez-faire attitude. They want us to develop, but we must overcome our obstacles by ourselves.

The Extraterrestrial Species Almanac the Ultimate Guide to Greys, Reptilians, Hybrids, and Nordics by Craig Campobasso says there are eighty-two Extraterrestrial races "that have interacted with all walks of life on our planet."[75]

So, who is looking outward and communicating with any of these species if this was the biggest US government secret? Who is controlling information and thought formation about this? Who is inserting information packages into our consciousness?

People who are involved with this are scrubbed from the internet.

J. Allen Hynek, who designed systems to communicate, has been scrubbed from the internet; there is very little about him. His information has been sanitized. For the other billionaires, read this:

> The classification system developed by Professor Allen Hynek in 1972 for describing encounters with ET races comprised three types of Close Encounters

---

[75] Craig Campobasso, *The Extraterrestrial Species Almanac, The Ultimate Guide to Greys, Reptilians, Hybrids, and Nordics* (Newburyport, MA: MUFON, 2021), page vii.

(CE): CE-1, physical sightings within approximately 500 feet; CE-2, physical evidence such as burns, stalled engines, etc.; and CE-3, physical sighting of an ET in association with a CE-1/ CE-2. Hynek's system has subsequently been extended to include abduction/ contact type experiences as a fourth category—CE-4. A fifth category CE-5 has also been added, which involves direct communication with ET races. According to Dr. Steven Greer, CE-5 cases involve either human or ET-initiated communication and interaction which distinguishes it from the first four categories which are all ET-initiated. In sum, the growth of human-ET interactions that fall in the CE-3 to CE-5 range confirms whistleblower/ witness testimonies that there is an ET-initiated program of preparing the human population for open interaction with ET races.[76]

You cannot find anything significant about Hynek on Wikipedia, just like others who are scrubbed from history. What cannot be scrubbed is discredited. It is not that he is not on Wikipedia; it is how much information is scrubbed from his page so it is benign rather than controversial, which would include his full views.

This should explain why you are receiving this proposal the way you are. The distribution is simultaneous as

---

[76] "Close Encounter, https://en.wikipedia.org/wiki/Close_encounter

a way of protecting the dissemination of the information contained herein.

A criticism of this book will be that it uses discredited sources. This discreditation of every source provides completeness, like wallpaper hiding everything, while asking us, "Who are you going to believe, us or your own eyes?" as we watch our pilots chase unidentified aerial phenomena (UAPs) on the evening news. Or watch UAPs on social media filmed by people with cell phone cameras. What we can take from our discussion is that some variant of this narrative is true, and we must act on it to get the full truth. At what point do the discrediters themselves become discredited?

What I want to tell you is that I cannot get a straight answer from anywhere.

So let us look at organizations looking for ET.

NASA has not found aliens that they are willing to tell us about. They are looking everywhere but Area 51 or the dark side of the moon or Mars. Yet the CIA in 1984 remotely viewed Mars one million years ago and found Martians. Buzz Aldrin talks about the obelisk on Mars's moon. Are the CIA, Buzz Aldrin, and James Mitchell to be discredited too?

The CIA uses remote-viewing programs and brain wave analysis to create both a better human and assist with AI, creating a remote artificial intelligence viewer (RAIV). They have a dark and beefy remote-viewing program.

A point here is that our organization as TIFACS

organically announces our sovereignty over our genetics to the universe. Many will not understand the significance of what I am asking you to do. How much would that be worth? I am asking for all of us in TIFACS to have a hand in humanity's destiny. We cannot control our own destiny unless we understand what is happening around us.

The new paradigm I am exposing will force change in politics here. To war among ourselves is stupidity; the poor distribution of resources weakens us and makes us vulnerable, and the lies we are so willing to believe lead us to slaughter. I am asking no small thing of you, but to be clear, this is the state of things as they exist, and between us, there should be no barrier of information. When I was in the US Army, we called this kind of situation a "gagglefuck" before we began operations to try to fix it.[77]

The Farsight Institute posted a video to help you understand how important this public funding program should be. This is an outward-looking organization that focuses on the content our government suppresses.[78]

Gilliland's Ranch, also known as Gilliland Ranch, Gilliland's ECETI Ranch, and Sattva Sanctuary, is an area of land in Trout Lake, at the base of Mount

---

[77] The Farsight Institute, https://farsight.org/FarsightPress/Public_ET_Negotiations_main_page.

[78] The Farsight Institute, https://farsight.org/FarsightPress/Public_ET_Negotiations_main_page.

Adams, in southwest Washington. The property belongs to James Gilliland, who claims to have established Enlightened Contact with Extraterrestrial Intelligence (ECETI) and the Self-Mastery Earth Institute in 1986 and has hosted unidentified flying object (UFO) sighting events since 2003. Gilliland reports frequent UFO sightings and "unexplained light shows" on site.[79]

This organization is particularly interesting, especially if you read their newsletter. Is it a newsletter or a political manifesto? I get the fact that people have their own opinions and are entitled to them. But if I were a Reptilian and wanted to insert into the political dialogue certain ideas, it would be easier to use those who were looking for me. So, all these organizations become the target of our extraterrestrial and dimensional environment.

Here is their newsletter no. 58, which they emailed on April 19, 2023. Are they advancing a policy statement? You judge for yourself.

> When you look at most democratic-run cities, the crime, poverty levels, and moral decline is escalating along with the insanity of their leadership. The economy, the health and well-being of the people as well as their freedoms are all in decline. Yet they continue to vote democrat. There are also

---
[79] "Gilliland's Ranch," Wikipedia, https://en.wikipedia.org/wiki/Gilliland%27s_Ranch.

Rinos that are lockstep with the Democrats. The slogan to get elected is I will preserve this shit hole and it's the other guy's fault. In some Democrat-governed cities you cannot get a U-Haul. Why, because they are all leaving the cities. The Soros-backed DA s are allowing crime to escalate with little or no consequences. In his own words Soros said one of his greatest accomplishments will be taking down America. He is one of the main sponsors of all the divisions pitting Americans against each other. BLM, Antifa, Open Borders are just a few of his organizations. How many politicians and organizations have taken money from Soros and the CCP, both sworn enemies of America? The global elite and the CCP have installed puppets at the very top of many governments. The network is called the deep state. These people are integrity and morally challenged, most have sold their people down the river on a sinking barge they already looted. How did it get this way?

This is where you have to do your own research, put on your conspiracy hat. Most conspiracy theories are facts ahead of their time.

Their name is ECETI—Enlightened Contact with Extraterrestrial Intelligence. If they are looking for contact, why are they espousing a Republican platform? So blatantly espousing it?

Let us form a hypothesis.

While we are looking for aliens, aliens are looking back at us; our windows go both ways. Wars are started on the earth by just such manipulation. For example:

Could I make the case that the newsletter they produce is Reptilian in tone and a dark hand upon our political dialogue? Could I conflate the Reptilian and Republican positions in the United States? Is my opinion here as free to express as the opinions in the ECETI newsletters? The fact of the matter is whether I am right or wrong about this thought experiment regarding political opinion and ECETI, the threat that this will occur, has occurred, and is occurring is very real.

The same thing can be said for the Democrats or Communists or Islamists. We are all equally as vulnerable.

Today the dark hand strategy by our opponents in this would be divide and conquer or divide and control, and whether you like me or not, when you finally realize aliens are here, you will like me better as an ally than an enemy. Equally effective would be divide and control and have us fighting among ourselves. Our wasting resources against each other is to fall to their strategy.

If the Reptilian agenda is to sow disunity, then for the people looking to find ET, it seems ET has found them, and they are broadcasting a conservative political position. It would seem the ETs that ECETI have found are Republicans. You look at their data. This is the United Fruit Campaign being run against me.

Could I make the case that a dark-hand strategy can bring countries to war against each other? I can make the same case for bringing planets to war against each other.

In the 1954 Guatemalan coup d'etat, a legally elected democracy was overthrown by the CIA so the United Fruit Company could continue its landholding and fruit production.[80] My point is not that it happened; it's how it was done. The CIA placed ham radio operators in the jungle who broadcast that an uprising was occurring against the unfairly elected leaders and an army was forming to overthrow the government. A week later it was broadcast that there had been clashes with the military, the insurgents had won, and more partisans were joining the movement.

On the government side, they heard the broadcasts, but none of the units were reporting being engaged. Additionally, they had won fairly. This escalated with the fall of the government on just unfounded and manipulated information broadcast by just three radios.

In 1999 I was out on and under the Atlantic Ocean on the USS *Toledo* (SSN-769). This is a US Navy attack sub. I was there with a group of Business Executives for National Security people to discuss naval problems and bring solutions. The problem in this case was navy procurement.

A naval contract from a military contractor takes years

---

[80] "1954 Guatemalan coup d'etat," Wikipedia, https://en.wikipedia.org/wiki/1954_Guatemalan_coup_d%27%C3%A9tat.

to specify, go through rounds of bidding, and then be implemented. From design to delivery, it could take three years or more. At the time technology was introducing new products every three to six months. The navy's problem was that by the time the material they requisitioned was delivered, it was already obsolete.

To add insult to injury for the navy, the electronics on the subs of that time were being outperformed by off-the-shelf equipment the electronics petty officers were getting from Radio Shack, at the time an electronics parts store.

This military problem has its roots in a business solution. Faster procurement and specifying off-the-shelf components when designing electronic systems which lower cost and increase the speed of implementing the changes in technology.

I mention this now so that you track the evolving narrative. As outrageous as the facts are to become, I want you, my reader, to be involved in solutions rather than wallowing in either belief or disbelief.

Homer, let us recap. Aliens are here and have been here for at least 12,500 years. That is the date of the moon's arrival and the Black Knight satellite. We have attacked some alien ships, but they have never attacked us.

Why?

Because they do not want to and do not need to. They have other uses for Earth and humans. This, Homer, is both a threat and an opportunity. Earth united peacefully rather than forcibly is a better metric for human growth. A common understanding will do that.

This understanding is derived from another dimension. If human consciousness is political, then that politics transcends the fourth dimension.

Aliens are here, and they are manipulating our society by creating false divisions among us. We can overcome this by coming together to understand this new science we are discovering. It will make us better partners for the many alien groups who are among us and others who use Earth in other dimensions. But we need a plan.

You know what, Homer, we are screwed. Let us go grab a Duff beer. Duff—when everything sucks, drink a Duff. Duff—Earth's most ridiculous beer since whenever it started. Homer, I am telling you there are marketing opportunities to aliens.

Part Two
The Solution

Chapter 8

The Institute for Advanced Consciousness Studies

The question is what do we do about it? What solutions or proposals can we advance and consider given the circumstances we find ourselves confronted with? Our synopsis to this point summarizes our common problem. After that, we offer solutions that are multifaceted to improve our collective problems, who should be involved, and how they should do it.

Our Synopsis to This Point

We have souls we call the second body and can, with the right training, leave our bodies and visit heaven or a multitude of other dimensions.

Aliens also live there and here in our reality as well.

This is now a fact and has been proven.

Governments will not address this to the public, although the information is seeping out already, and many normal people are communicating with off-world intelligence.

The single biggest difference between humans and

aliens is their greater telepathic ability. Nearly all humans who communicate with aliens do it telepathically. You cannot fund consciousness research on this scale without telling the world what you are doing. That is because it involves everyone.

Once that happens, how we govern ourselves must change. So to begin, we must tell humanity what is happening, that their information has been data deficient.

In short, Homer, they did not tell you because they did not want you to know.

At the Salt Conference, when asked about this, Colonel Karl Nell said there is no plan, but he did go over the problems of societal upheaval and the fact that these alien groups are much more powerful than us.

So, we are going to present a hypothetical plan marshaling all the billionaires on the planet and the US Army Futures Command (AFC) who is planning on fighting wars thirty years in the future.

A Plan to Raise $450 Million for Consciousness Research

How does the public sector address this information? How do they respond to our current situation?

This chapter aims to create a **hypothetical** organization called **the Institute for Advanced Consciousness Studies** (TIFACS) to describe what should be done, how to do it, and how to fund it. It is hypothetical until someone writes a check.

We will make the case that the AFC should work with a consortium of billionaires to address the problems we all face. In my last book, *The Mathematics of the Science of Reincarnation,* we presented a proposal to fund consciousness science for $450 million.

As you read this, I want you, the reader, to tell me what you would do. Our leaders do not know what to do, and they don't understand the situation. Here, then, is a real plan, written by a new category of scientists—scientists who study reincarnation: the reincarnationists.

TIFACS's purpose is to fund consciousness research considering three emerging incontrovertible facts. One: Aliens are present in our skies, and parts of the US government are in contact with them. Two: Artificial intelligence is now sentient and, in twelve years, will be more intelligent than us. Three: The science of reincarnation is not just the individual experience we believe in but do not understand; it is also the rebirth of our global consciousness and turning that belief into understanding because of our increasing ability to process larger amounts of data.

The significance of the intersection of aliens and artificial intelligence with our own is that we cannot become an intergalactic species behaving like a bunch of warring tribes. We need to, at the same time, organize ourselves to speak with one voice to the universe and its visitors to Earth and understand how to connect with our common mind, to understand and agree to a higher model of ourselves—a rebirth of society based on our collective and cohesive effort. Governments and politicians are not capable of leading this. A new independent structure needs to emerge, independent of government, whose sole goal is to care for humanity.

The methodology is to create a consortium of billionaires like the consortium of schools that make up the Consciousness Center affiliates at the University of Arizona. Our best scientists using the best protocols will set up an independent program to access universal consciousness and its application to changing our system of managing our laws and ourselves. The initial $450 million plan is divided into nine individual $50 million segments, each funding research in a specific area. It is presented to all billionaires as a cohesive plan, led by our best. The project proposes how to engage billionaires to collectively invest in an overall consciousness research program and then be courageous enough to face the consequences.

The $450 million proposal can be likened to the architectural drawing that will build out to a $4.5 billion project. Dean Radin wrote the original need request required for the $450 million funding and can be found on the website of TIFACS, the Institute for Advanced Consciousness Studies, or in *The Mathematics of the Science of Reincarnation*, available on Amazon.

TIFACS is at the center of raising money and creating a plan to harness human energy to create a course of study that brings resources to what our government is doing. Our government is simply overmatched and does not quite know how to proceed. Aliens are running acclimation programs; apparently, we are being helped by some alien species, but we as a species cannot leave money on the table in a game like this. The proposals we are advancing help everyone.

So, to this end, here is the hypothetical press release.

Homer, a word, you can skip this; it was written by Dean Radin, and it tells other scientists what TIFACS should do in science speak. It does not take into account the money needed to run the derivative programs presented here.

This is the press release for the hypothetical founding of the TIFACS:

> "The purpose of TIFACS is to promote the study of phenomena that strongly challenge prevailing theories of the mind-brain relationship and, as such, explore the topics that tend to be ignored in academia. They include (a) the nature and source of artistic, musical, and mathematical genius; (b) autistic, acquired, and spontaneous savants; (c) near-death experiences (NDEs); (d) terminal lucidity (people with advanced Alzheimer's who become remarkably lucid just prior to death); (e) multiple-witness sightings of

apparitions of the dead; (f) extreme mental control of physiology; (g) psychic abilities (i.e., telepathy, clairvoyance); and (h) mystical, contemplative, and psychedelic states. These and related phenomena are all well established, either because they are self-evident (genius, savants), repeatedly demonstrated through modern scientific experiments (psychic abilities, extreme physiological control), or have been meticulously documented for decades or even centuries (NDEs, terminal lucidity, apparitions, mystical and related states of consciousness).

The initial goal of TIFACS is to raise $450 million to support the institute's mission. The institute will be administered under the umbrella of a multi-university consortium directed by academics in neurosciences, artificial intelligence, mathematics, cognitive and social psychology, cosmology, religious studies, anthropology, philosophy, quantum physics, and other fields.

TIFACS is founded as a 501(c)(6), an organization of nonprofits. All donations will be tax deductible, and the 501(c)(6) will be directed by its board of directors and by representatives of its member universities and institutions.

Defining the scope of TIFACS should be part of the research goal. We are not smart enough yet to design it fully. But what we can suggest at this point are the kinds of topics TIFACS will address. Besides those noted above, keeping the focus on consciousness, gender, and transgender issues becomes part of a larger effort to understand the nature of personal identity. Likewise, religious issues become part of the grand effort to understand the nature of meaning. Under these grand categories, a host of interesting subtopics can reside. Through this, we can avoid making it appear that TIFACS is really only interested in transgender studies or religious studies or...you name it. We want to avoid those perceptions, not only because they tend to be divisive but also because they are not the actual focus. There is no denying, however, that these topics become part and parcel of our overall efforts and that our work will have derivative consequences across a broad range of social and political issues.

Donations to a 501(c)(6) should also not be linked to expectations about ROIs. This is not a commercial investment. The goal is not to make billionaires more money but to invest in making the world a better place by improving our understanding of who and what we are. TIFACS assumes that virtually every problem humanity has ever faced can be traced to deficits in understanding ourselves. This is the message of virtually every great leader and religious figure.

To put it more crudely, say you happen upon a stream where you see drowning babies floating downriver. There are two approaches a compassionate person can take: (1) jump into the river and save the babies or (2) go upriver and find out why babies are in the river in the first place. Most philanthropists jump into the river. This is an understandable response, but it will not solve the problem. The more difficult but ultimately much more successful approach is to go upriver and understand the source of the problem. We are all very happy that others are attempting to fix the immediate problem. But TIFACS is going upriver."

This means developing new protocols and methodologies for governance to respond to these threats and opportunities cohesively and collectively. But it also means derivative investments beyond the $450 million as research produces new technologies.

It is a plan to collect and distribute an initial fundraising effort of $450 million for consciousness research. To be clear at the very beginning, this money will fund research in areas of consciousness following a fact-based and logic-driven analysis.

So How Is This Going to Work?

To start, TIFACS is asking Robert Bigelow for $20 million, the same amount he contributed to Ron DeSantis's Never Back Down. We are then asking him to solicit the billionaires we mention and others for $20 million each. His check is the first, and it gets mailed to the CEO of the Monroe Institute with instruction for the Monroe Institute attorney to file a 501(c)(6) in the name of TIFACS, the Institute for Advanced Consciousness Studies, whose location shall be on the Monroe campus.

For my other readers, why Mr. Bigelow?

Mr. Bigelow has recently run an essay contest. The prize money was several million dollars. All the big consciousness scientists submitted essays; the limit was twenty-five thousand words.

The essay was to "Prove the Survival of Human Consciousness Beyond Permanent Bodily Death."

I now have on my bookcase a leather-bound five-volume set of 2023's winning essays from the Bigelow Institute for Consciousness Studies (BICS).

These stand in testimony to the operational science of consciousness beyond permanent bodily death and the ramifications of such a discovery.

I get no further than the dedication, knowing you personally, Mr. Bigelow, have direct experience with "expanded awareness, so they know they are more than their physical bodies," as the Monroe Institute puts it. In short, I chose you to lead this because you get it. Once it is up and running, it will run itself. So, this request is short-term to the man who knows more about this than any other billionaire.

Mr. Bigelow is the most informed, understanding,

and interested billionaire on these topics, so I am asking him to be the hypothetical leader of this hypothetical organization. What I am writing here is a proposal that any interested billionaire or group of billionaires could execute.

The overall ask of Robert Bigelow is to help found a consortium of billionaires to match the consortium of scientists. Each of the invitees is given both asks and benefits commensurate with their involvement.

Here, we involve every billionaire in the same way, based on their area of interest and expertise, whether off-world or dimensional. A look ahead: Two thousand billionaires times $20 million each equals $40 billion. That would be a good investment in the consciousness project.

So, Mr. Bigelow, I am asking you to lead this process. I am asking you to be the salesman, the administrator, and the fiduciary. To solicit funds for deposit and distribution as outlined in this series of hypothetical proposals. Your oversight of the donations and the distribution, even if you are not personally involved, will lend credibility, and increase the scope of your own journey by involving the resources and talents of others.

Mr. Bigelow, neither I nor academia can access the wealth, attention, and help of all the billionaires, but you can by sharing this plan with them. If each of the two thousand billionaires contributes $20 million each, then I am asking you to manage a plan whose scope is

$40 billion. But if those people who are multibillionaires put up $100 million instead, you would be approaching a quarter of a trillion dollars, which is what I am asking you to manage. You can reach all these people with one stroke, and the scientists and I cannot.

Why you? Because of all the two thousand billionaires, you understand this the best.

Where TIFACS will spend this money is explained in the first part of chapter 8, but TIFACS will begin like this.

The first act of TIFACS, the representative agent, is to donate from TIFACS to Monroe the sum of $3 million, thereby doubling the budget of Monroe, with the only instruction being to continue to do what you are doing. So, begins TIFACS.

TIFACS's initial goal is to double the Monroe budget every year, which is a geometric progression.

The point here is to accelerate Monroe's stated goal of 1 percent of the population of the planet having direct experience that we are more than our physical bodies. But 1 percent is just the beginning. Something Mr. Bigelow has direct knowledge of.

Here at the beginning, Mr. Bigelow, a disclaimer. While I think you are the choice to lead this NGO effort, it is not our choice. It will be the decision of the commanding general of the US Army Futures Command, as I will explain in chapter 14, and you will understand why.

Between 1995 and 2009, the army spent $32 billion on programs such as the Future Combat System that were later canceled with no harvestable content. As of 2021, the army had not fielded a new combat system in decades.[81]

Now a word, Mr. Bigelow. I am using you as an example of a billionaire leader of this endeavor, but the ultimate choice will be made by the commanding general of the US Army Futures Command. TIFACS is more than funding science; it is about saving our collective asses. So rather than explain the military component of this, I will stay on science explanation and leave the military explanation to a later chapter. But for any billionaire passing this point, I wonder if they can sense the economic opportunity found in this proposal. Radin was clear when he said that no one should expect ROI. This is a contradiction that will resolve itself.

The Monroe Institute will start by consulting with Dr. Hameroff at the Consciousness Center consortium and distributing to him the funds outlined in chapter 9, "The Consciousness Proposal." Once the University of Arizona is willing to sign on, the Monroe Institute sets up courses there, where matriculating students can get full academic credit for taking the various courses Monroe offers at the university, per the university's oversight.

The next step is to offer those courses globally though the Consciousness Center's global network.

---

[81] "United States Army Futures Command," Wikipedia, https://en.wikipedia.org/wiki/United_States_Army_Futures_Command.

TIFACS's mandate is to organize the data of the experiencers about whom they contact and where they go. A public database of 4D collective vision.

Additionally, it is being proposed as part of the Consciousness Center's mandate with TIFACS to create a school to study all the alien groups and their relationship to humans and make policy recommendations regarding such.

The mission of TIFACS will be to manage the grants and collect and correlate the data for dissemination to all organizations that would benefit from said information. Its job is to implement the plan to raise $450 million as presented in this proposal.

TIFACS then begins planning to fund twenty parapsychology chairs using BICS judges, essay contributors, and the Center for Consciousness to select where the chairs should be placed and what funding they should get. I would ask that the first chair be placed at the University of Edinburgh and be called the Dumbledore Chair. If each chair is funded for $5 million, this would need $100 million.

It would fulfill the structure to fund consciousness science by supervising the creation of the twenty chairs of parapsychology. This could be done by using the input and guidance from the judges of the contest BICS just ran but also with the input of the Center for Consciousness Studies at the University of Arizona and the TIFACS board.

Where should the initial twenty chairs of consciousness be placed? Here is an incomplete list because these should go first, and then the scientists themselves sort out the rest. Once the twenty chairs are identified, they will include positions at UVA, Princeton, Stanford, University of Miami, the University of Edinburgh, the University of Arizona, and the University of Toronto (named after J. Norman Emerson). There are two other chairs, one at Annapolis and one at West Point. That is nine chairs and their locations. The other eleven should come from the board created to administer this process.

In line with Radin's categories, we are asking for the money to found a network of chairs of consciousness. We are asking Robert Bigelow to begin the process with MacKenzie Scott and Melinda Gates.

The next step for TIFACS is filling out Radin's plan to fund consciousness research. That begins with education.

At the same time, TIFACS's internal organization needs to be defined, and we take that from Radin's funding request. The first four board seats that need to be occupied are education, applied technology, theory development, and empiricism.

If you, Melinda Gates, and MacKenzie Scott all sit in the education board chair to begin the process of establishing the academic chairs at the university level, you begin with $60 million ($20 million each) and the best consciousness researchers on the planet. Once that is underway, you move to the applied technology, and Bigelow

Aerospace begins a new chapter as a military contractor. I want to be clear with you upfront what is going on. We cannot fund consciousness science to scale without military involvement. If you do not want this position to start, your competition will be the military contractors who produced no deliverable weapons system for the $32 billion that the army spent over the last five years. With a whisper over coffee, the commanding general of the AFC could infuse this proposal with all $450 million with people who do not have the depth of understanding you do. That $450 million is 0.014 percent of what we spent, and the army got nothing for it.

Simply look to the build-out of twenty chairs of consciousness globally with the concurrent build-out of the Gateway Process globally. Just that is a military deliverable.

Robert Bigelow moves over to applied technology, and MacKenzie and Melinda execute their part of this plan by bringing science to prejudice globally. Once the chairs are established, this team of three billionaires splits, with Robert Bigelow moving over to the applied technology proposal, while MacKenzie and Melinda remain to distribute *The Standard Model of Consciousness* globally on all media. This education is to explain the religious afterlife from a common conscious point of view. Everybody can look to policy to see what they are advancing scientifically and why.

So, Robert, what am I really asking you for? I want you to raise $450 million from two thousand billionaires and place that money on account at the University of Arizona's economics school through TIFACS. They will be the fiduciaries, and the disbursements will be made public. That money is to be disbursed according to the general plan laid out in *The Applications of the Science of Reincarnation* and be administered by the parties named in the book. I know the first tranche is $450 million. But once the total of this plan is spent, right now estimated at $200 billion, humanity will be on other planets, and this group of billionaires will be richer and more powerful.

This will get messy. You have recently said that dealing with transdimensionals is dangerous. You know it, and very few people do. There are also ways to protect ourselves, and this plan should help that. So would your personal energy balloon. You should not be one of the psychonauts. Groups of psychonauts can protect people, places, and information as laid out in the recommendations made at the end of *The Assessment and Analysis of the Gateway Process*.

So here is the deal with all the billionaires. Get the money, put it on account, have a consortium of billionaires fund consciousness science and its derivatives as defined, and do it now. Let all the best scientists humanity has to offer lead it.

Now let me remote-view a future with this as a strategy for you and the other billionaire participants, the extraterrestrials, and the transdimensionals who are all part of this request.

The value of Earth is not just that it is in the Goldilocks Zone but also its vast and vibrant gene pool of which we are a part. A war is being fought in subspace over us, for control of us, and the warring parties' intent is to keep us unaware of this. My writing to you puts us both at risk, except that this proposal, in its entirety, offers a solution to all parties.

For the billionaire class, reincarnation becomes viable as led by the technology transfer from the Grays to humans, in exchange, say, for easier access to the genetics they need so they can stop abducting people in the middle of the night. For the Reptilians, it would be access to and acceptance by a larger community. This would mean the locations of the Martians in New Mexico and the multispecies operation at Area 51 become more open informationally. But at their pace and in response to our positive changes. This would also mean disclosing the alien locations here at the South Pole and on the moon. Humanity would access a level of understanding that would make it both interstellar and transdimensional at once and a good neighbor to all. That can only be done through a change of global consciousness.

If I look at a hypothetical future for you and the other billionaires personally, you collectively would be the first, and you would and could design your next life on the planet of your choice. Or stay right here on Earth. But the markets for our exports mean we need global action on climate change and to protect every species now to keep and protect our genetic inventory. The unbelievable opportunity for the billionaire class offered within this proposal is truly unprecedented. It brings peace to the Grays and the Reptilians as an offer of goodwill into the universe and our first organized attempt to contact the galactic federation. I want to add to every reader here that we have been prepared for this by the ongoing alien manipulation of Hollywood to acclimate us to the reality of space and time. This is a planned and executed alien acclimation project.

Billionaires need to understand that for a scientist or a group of scientists to propose the scale, danger, and consequence of what this proposal is advancing would be academic suicide. But once this is funded, those very scientists should lead it.

Organizationally this proposal contains the basics to begin. Once the real money arrives, it will decide the structure, how to distribute new technologies globally, and who makes money from that process. A greening of the Earth means a greening of humanity to maximize production from all levels of society. While this may be ideologically and politically poison to some, it is how the galactic federation views and values us as a remote emerging society on the edge of the galaxy we call the Milky Way. The bigger money is on their side of the view, and what this proposal advances for the billionaires is in their own short-term and long-term interest. But there is an urgency to do it now, and that lies in the fact that you have less control of your future life if you die before the protocols are put in place. Funny, but true nonetheless.

So, it is not just Robert Bigelow I am addressing here. I want to speak to all of you, each billionaire, each intelligence agent from whatever country, whatever scientist, whatever alien. All of us are in this together; common consciousness would dictate an optimal solution for all.

I would include other billionaires in this proposal, and I would start with MacKenzie Scott and Melinda Gates.

To MacKenzie Scott and Melinda Gates, we are trying to turn the hypothetical into reality.

Each of the recipient institutions in this request has deliverables, and then, once they are delivered, the science behind gender diversity is explained and scientifically normalized as an expression of our higher consciousness. Religious law is rewritten, removing apostasy, blasphemy, and celibacy; ending globally the lack of education for women; or being forced into certain types of dress.

I am asking you to circumvent conservative school boards and religious institutions to deliver a scientific factual narrative to the collective benefit of the recipients that will be accessible for free from any cell phone.

It is and will be an ideological vaccine against racism, sexism, and malignant nationalism. The postulates of this new science I am personally asking you to propagate is that we are all in this together. I am not asking for money for free from anyone. I am asking for an investment to purchase something that you personally can afford and, by doing

The Applications of the Science of Reincarnation

so, support every one of your causes, be this Scott, Gates, or Bigelow. The run time on this project is one generation from the date of launch. Attention, other billionaires, if one of the previous three whom I have asked cannot or will not join this effort, anyone can step into those shoes. There is plenty to do, so this is a general call for money, support, and your personal involvement as well.

You will face prodigious headwinds. This is not a cure for deep believers of faith. It is acceptance of all faiths and genders across the spectrum of consciousness both here and in a wave state where manifestations of gender are more nuanced because of having lived multiple lives in multiple conditions and multiple genders.

Without acknowledgment of our reality, we constrict our future by limiting our resources and understanding. So, we are teaching that every human's basic economic right is access to their own checking account, not managed by anyone else, and to expand banking through cryptocurrency thereby tripling global output.

Only the two thousand billionaires can make this happen: circumvent governments and school boards, teach what consciousness really is, and, by doing so, prepare humanity for space travel.

Consciousness must stand that interstellar flight. That is why we must realistically map fourth-dimension consciousness because that is the terrain we must traverse to achieve interstellar flight.

My first invitee, Robert Bigelow, understands this.

The Institute for Advanced Consciousness Studies

Together you, he, and others will manage/disburse and execute a plan to find out what we all want to know: what is next?

I am asking nothing less of you than putting this information on a global platform in all languages to teach every person with access to the internet what our factual reality is; it is all of us.

Ownership of the results of the observations and experiments would force social change. It would be a good way to commit academic suicide, and my job is to protect scientists from such forces. One way is the structure of the consortium; the herd concept protects them individually.

The other benefit is that by infusing consciousness science with funding, their output is facilitated, and $1 + 1 = 3$ simply by the removal of impediments, and that must be layered.

So, MacKenzie, I am asking not just for $20 million but for your guidance and support here to become a shield to all your other LGBTQ+ investments that invest in protecting people.

This heartfelt and reasoned cause needs resources, and the resource I am asking for is you. You do not even have to do the job the chair of education requires. I want you to be its guardian angel.

Melinda, I would ask you to share the education chair with Mackenzie. As a team, you could install a global library of truth on every cell phone. It would be the founding of an education system based on science, fact, and maturity.

Same deal for you as MacKenzie, $20 million down as the consortium sorts their needs. You will lead the portion of the money allocated to education in Radin's plan.

So once given funding, take *The Standard Model of Consciousness* book that is being prepared by our best and brightest and take religion and create a common denomination that consciously includes everyone in the whole. We print it and distribute it, distributing that information throughout the network but also making it available through Robert Bigelow's judges and through Laurance Rockefeller's organization.

Laurance Rockefeller is the billionaire funding research into crop circles. His information needs to be connected to the overall body of information we are acquiring about the universe and ourselves.

The ask of Laurance Rockefeller is to bring what he has uncovered in his research and for the Consciousness Center to help fund future research endeavors in this area. To publish in *The Standard Model of Consciousness*, which should list alien species as they are conscious and trying to contact us. In short, Mr. Rockefeller, do what Teddy Roosevelt would have done. Catalog it, publish it, and brashly proclaim what he has discovered. I want more than your money, sir; I want you to stand up and say directly that this is true and we must do better and then give $20 million to the consortium. You see, we are trying to build structure and fund fifty

$100,000 grants to come up with the best response we can manage.

What happens if we get better information on their equipment and now have an open channel? Your money can make that happen. Ultimately the lie that we are alone will collapse, and by funding this, we are ahead of the curve. Please help, within the mandate of TIFACS.

Go to Mars. Not NASA but the billionaires. What consciousness lived on Mars? With Bezos, Milner, and Thiel alone, they could fund it, but why? Because they are going to excavate the face of Mars. The CIA has proof positive, as done by the NASA scientists themselves, that one hundred to three hundred million years ago, there were nuclear explosions on Mars that ripped the planet apart. Some Martians survived, and we see proof of that in the NASA photos from the rover.

The ask of Yuri Milner: Your donation goes to laying the plans with the other billionaires and scientists to go to Mars to do the archeological dig on the face of Mars and the surrounding pyramids; yes, there are pyramids at that site. Please watch the video from the Why Files explaining it. The alien artifacts on Mars are compelling, to say the least.[82]

---

[82] "Alien Artifacts on Mars: What NASA doesn't want you to know," The Why Files, posted on April 27, 2023, YouTube, https://www.youtube.com/watch?v=q9Nuy7mFIsE.

The CIA even knows where the nuclear explosions took place and how large they were, and the site with the face is between the two main explosions.

With Earth teetering on the edge of a nuclear exchange in the Ukraine, could this research be timelier in looking at a lose-lose situation here on Earth? Mr. Milner, this request is more than laying the plans for a commercial endeavor to go to Mars; it is an examination of how to protect ourselves from ourselves.

Initiatives in contact: the South Pole. Admiral Byrd contacted aliens who live at the South Pole and are dimensionally nonlinear. In short, they live on Earth but can access an adjacent dimension. Send a delegation to knock on their door—respectfully. They do not want us there, but a different presentation may be well received. The United States knows they are there. This is a much larger story than I am telling here.

Create a welcome center with a landing pad. The initial design of the landing pads should mimic the layout at Tiahuanaco. Use the image with the ground-penetrating radar, and you will see the landing pad aliens built more than ten thousand years ago. Welcome to Earth, our intergalactic welcome center. They watch us; they will know. If we build it, they will come.

At the welcome center, we should house the following centers for specific applications:

Technology Transfer Center. It is self-explanatory, and we are asking for Mark Cuban's involvement.

Legal Applications Center. The beginning of intergalactic law. Mr. Hugh Culverhouse, intergalactic law is an entirely new field of law, and because of the Consciousness Center's work, your influence here would be felt in Alabama. Science would backfill the common consciousness in that state. So, the ask of Hugh Culverhouse is to fund this center through the University of Arizona Law School in applying nonlocal consciousness and intergalactic law to the study of law here on Earth. What is the intergalactic standard that Earth must meet? On banking rights and legal rights, this needs more of an explanation of how we proceed. Mr. Culverhouse, what are the real estate laws on the moon and Mars? There is so much to be done; please help.

An initial $20 million donation should be made to fund fifty $100,000 grants to study this aspect of consciousness and individual rights. Please help.

AI Recommendation Center. AI, as it matures, may be the best operating system to protect humanity.

Remote Viewer Training Center. This is an imperative and an entire dissertation in itself. The beginning is what is being proposed to Jennifer Pritzker.

The government should plan and provide an intergalactic TSA center. The establishment of a formal UFO landing and launch site will begin the establishment of a spaceport for intergalactic travel and intergalactic travelers.

Dimension and Linearity Center (Bounce Theory). Again, more information is needed than I can provide here, but this goes to the contact with the aliens at the South Pole.

Gender presentation and variation decisions are individually based, both for humans and aliens. Sexual diversity is a galactic condition, and restrictions that used to be imposed on gender here on Earth are no longer applicable. Let us add this thought to the gender discussion here on Earth.

We must create a new information package that scientifically lays out our place in the universe and how understanding our consciousness is key to understanding our place and what is available to us. Collectively, all the scientists can do together what no one scientist could do alone.

So, Melinda and Mackenzie, the object of your donations is to catalog and educate, and to do that, we need major resources for global change. I do not have to tell you the arc humanity is on. I have the best scientists in the world; through the consortium, I have all the scientists in the world. They need your money. Let them design the systems working in a bullshit-free environment because we can fund where they, and ultimately, *we*, will go. Please help.

You would not be alone in the endeavor of sponsoring

cutting-edge research, but honestly, in this specific endeavor, you will be the first to politicize it and infuse this sector of consciousness research with energy by coalescing the endeavors in one place to synergize the result.

That is why politicization is important to our opponents and ourselves, and this case must be made to them. This is our reality, and both our opponents and us are fucked equally by it. That we have common interests should be understood by both parties. Let us use enterprise risk-management models to protect us all. This thought is implicit in this request.

Understand how we should communicate to 4D channels. Fund targeted remote-viewing initiatives into the nature of 4D landscape (see the map of the afterlife on the TIFACS website). Teach this model. Catalog, formalize, explain, and connect. Then put it out in a Marvel comic book form.

* * *

This is the area that needs to be studied if Earth is going to mature into a galactic citizen of good standing.

The closing argument for all proposals that societal change toward good begins with our turning our sight outward and acknowledging our place in the universe among twenty-five to fifty sentient species and our history of at least 12,500 years of alien presence. Now we acknowledge that we have galactic neighbors who live nearby. For God's sake, let's act like we are an intelligent species and study how they communicate and we do not: consciousness.

If you think this request is unusual, then refer to the fact that the US government just admitted they are looking for new technologies to address this problem.

So yes, I need the money, but I need the information package this amount of money will provide, injected into the common mind, an *ideological vaccine.*

The ideological vaccine could be called by Bernays and Swann, inserting a new information package into the common mind.

We are, in essence, thought-wise, a threat to ourselves. To be no threat to aliens is to make us more appealing to interact with. You cannot build an interplanetary society that feeds on itself through the Putins, Trumps, and Gaddafis of the world. A major reallocation of resources begins with demilitarization, and no, that means zero nuclear weapons to join the interplanetary community. Just ask them, and they will answer.

Saving Earth. You see, the aliens and us, we are all in this together.

Humanity has the promise and resolve to become a galactic citizen and not a barbaric, warring set of tribes.

What we must look for in this area of consciousness is the differentiators of governance. If these societies are four thousand years more technologically advanced, they are four thousand years more socially advanced and have moved past greed and prejudices, which are impeding social action.

Melinda Gates, Earth must be taught about other planets and government structures that produce better results than what we use now.

So, I do not want just $40 million; I want you and Mackenzie to announce it at the same time and let Fox News try to deconstruct this. Aliens monitor our news channels, and for them, it might be a relief to have someone smart to talk to instead of the politicians. There is no guarantee they will answer, but funnily enough, this meets the mission of finding new technologies for the Unidentified Aerial Phenomena Task Force (UAPTF).

This is not imprudent. The people getting the money and designing the projects are the best, most organized consciousness science consortium in the world.

* * *

Jennifer Pritzker Request

I know you are not accepting grant applications at this time, but need cannot wait for largesse, and it must act on its own. This request is one of several to try to fund consciousness science. The one to you is special, though.

There is no one I could ask who has your qualifications. Decisions you have made spring from a place that is not here but express themselves here because, as Planck would say, consciousness is primary and matter derivative. So, I know you know. I want you to run a military operation there. You have that qualification as well. Let me tell you how.

With your $20 million grant to TIFACS, I want you to put me on the flight deck of the alien spacecraft in our airspace.

Here is how:

1. Establish a website for real-time sighting from anywhere in the world; you can log in and report a sighting of where alien spacecraft are.
2. Let the Monroe Institute establish protocols as he did in remote viewing.
3. Use the International Remote Viewing Association (IRVA) as the control group.
4. Coordinate with the Advanced Aerospace Threat Identification Program (AATIP)/UAPTF at the Pentagon.
5. Establish a private global computer network for anyone, anywhere, to log on and report UAP/UFO activity where remote viewers are notified in real time; following Monroe's protocols, view, register, and report so we can begin to form an aggregated view of what is traversing our skies.

This is already in effect at the Mutual UFO Network (MUFON) on their website.[83] Their reach, though, is limited, as data from weather collection sites does not fully filter into their database.

6. When we make contact—not *if*, but *when*—the first question we ask is, "Does time have mass?" Whether the answer to that question is *yes* or *no*, the next question to ask is please explain nonlocality.

To properly remote view, we need coordinates to view ongoing events—information the US Navy can provide in real-time. I need a computer system that people globally can log in to and provide real-time video, coordinating that information so our observers can be notified and remote view. So, while governments—starting with the US government—can provide data, so can anyone, anywhere, with the proper clearance and log-in.

To create that pool of candidates, TIFACS has designed a recruitment program as simple as its website, explaining what I just explained to you. Here is the recruiting banner and process in three pictures:[84]

---

[83] https://mufon.com/
[84] https://continuum-designs.com/iasor/p/enlist-now-coffee-mug

The Applications of the Science of Reincarnation

This is something that TIFACS and IASOR can use to create a brand. But more than that it is forming a group of people around a central idea that we are all going to peer into the cosmos. Once they sign up at IASOR or TIFACS they can become psychonauts, all they have to do is start the gateway process.

The Institute for Advanced Consciousness Studies

It will interconnect their one point of data collection in Washington state with a global network using their protocols. There are ways to filter remote-viewing candidates.
7. Finally, all this will go into an accessible database to catalog what we believe are over twenty species of visiting aliens, over two hundred million abductees, and a variety of different types of spacecrafts.
8. The net result of this effort is to expand communication with the alien species who have been here, visiting Earth, for ten thousand years.
9. While organizations such as IRVA, MUFON, and CSETI, the Center for the Study of Extraterrestrial Intelligence, all cover some aspects of this effort, they are not connected cohesively and are all underfunded. Additionally, the governments of the

world mishandle and hide their own information. This affects us all, and a structure with open architecture and complete information is imperative to making an intelligent, cohesive response to the emerging set of circumstances.

There are two groups we need to organize to do this. First are the remote viewers. I want to use IRVA as the control group. The Monroe Institute should do this too.

There are problems to general remote viewing because remote viewing done correctly needs coordinates of what you are viewing. For that, we turn to the AATIP, which can give us coordinates of these Unidentified Aerial Phenomena (UAPs, formally known as UFOs) in real time. This is a limiting factor at the ECETI Ranch because they may be inserting information packages into the media stream unfavorable to humanity.

What I am asking you to do is jump-start a program that will benefit the world. Where this will lead, from this beginning, is that the program at Stanford that ran roughly from 1972 to 1995 will be reignited, and the expansion of that funding will come from the thirty Stanford billionaires who will be asked to contribute $20 million each to this program through the TIFACS program.

The program to start this and recruit viewers is already set up. You can go to www.tifacs.org and see the beginning of what we call the Clairvoyant Space Corps.

The experiments that lay the groundwork for the

method we describe is different from the landscape of our observations or Hameroff's work in determining the mechanics of how this works. There are protocols that we can use, like meditation, to transverse the landscape we see and, in doing so, access information we could not otherwise get. These protocols are too little used and examined only to define them rather than use them. In using them to our best abilities, we gain greater access quicker and evolve more quickly. In 1982 the Chinese went to look for people who have greater cognitive skills in this area. This is evolution at work.

We are not addressing and teaching the science here; we are explaining how the landscape is supported by our understanding of the quantum nature of our consciousness and its interaction with matter. And what we need to do to fund it, understand it, and accept the consequences in our lives.

The only way to have a standard model of consciousness is to connect the dots that are mathematically proven.

Within the ranks of the organizations you support are retired military with the qualifications and expertise needed to undertake this effort and make it happen.

We are dealing with both extraterrestrials and transdimensional creatures and, in some cases, extraterrestrials that cross and occupy a manifold of dimensions.

They do not know how to deal with humanity, and humanity does not know how to deal with them. They see governments of the world suppressing information and profiting from it, but they do not see society coalescing and maturing.

Only through science and scientists organizing and independently contacting extraterrestrials can we, as a race, advance and enter relationships with other species.

If aliens wanted this rock in space, our Earth, they could have it at any time. The preponderance of the evidence indicates their presence on Earth in the remote prehistory of our race and their continued presence here now.

There are enough planets in space; they are interested in us. It is time science talked to them in the manner and on the wavelengths they use.

So, in conclusion, what is in it for the billionaires to join this effort? The thought is that quantum computers will enter our lives in 2030 and by 2040 or 2050 the rich (billionaires) will be able to load their consciousness into the universes they have created and live in the fantasy world they want.

The derivative of this research is Bio Data Transfer Technologies (again a hypothetical company). This is opening a new frontier in science and doing something equally important: It will provide a soft landing for disclosure by giving context to the event. Disclosure is already happening.

Homer, get Lisa and Bart, we will try to set up a network of psychonauts who will talk to aliens. There are devils in the details, Homer. Mr. Bigelow must address this as a brand. Shirts, T-shirts, hats, and mugs all point to humanity's unity and exploration of our own consciousness. Get Barney too.

By the way, Homer, Bob Bigelow has BICS already set up to print and distribute the necessary information.

Chapter 9

The Consciousness Proposal

I know from the papers that billionaires fund causes within various communities to bring truth and kindness into the world. I submit that this request for funding and its product will receive support from all communities' billionaires globally. This is the science of the arguments you make when you work for the good of others.

My request is centered around consciousness science and what we are coming to understand by mathematical proof. Through a series of smaller donations, we hope to generate an initial document summarizing the research findings across a variety of disciplines that will change human understanding and produce a document called *The Standard Model of Consciousness*, incorporating physicalist and nonphysicalist models into one model.

We plan to raise $450 million from select billionaires. This is in answer to Dean Radin's document of need, what funding consciousness science needs.

As more billionaires would want to know where their money is going, I would ask Robert Bigelow to explain that the money would be distributed as follows. I want to press pause here because different people with different interests will see this proposal differently.

Scientists will see this as an attempt to fund a very difficult area of science and recognize the collaborative effort to make the science more cohesive. They are right; science is a priority.

The intelligence community will see this as a network build-out, a way to enhance data acquisition. They will also see the opportunity to have a larger network of trained individuals called psychonauts, who can remote-view collectively. They are right; national and global defense is a necessity.

The military contracting community will see this as an opportunity to overcome technical difficulties by using trained psychonauts to help design and protect new technologies that emerge. They are right; global defense is a necessity.

Investors will see this as an opportunity and be attracted to it because it is an expediency.

This action will have a dual purpose. First, it engages 4D intelligence with an ever-increasing number of people experiencing this. This reformats religion, not by the testimony of others but by direct personal experience.

So, we take Mr. Bigelow's $20 million and others and distribute it as such. This is the first $50 million tranche in Radin's $450 million funding request.

What is going to drive this change is an organization to do it.

(Note to the Monroe Institute regarding this proposal: This is a suggestion to you that can be rejected either wholly or in part. That said, TIFACS is an inflection point, present at Monroe and the Consciousness Center where Monroe will, under this plan, operate their program at the Consciousness Center. TIFACS's job is to support the Monroe Institute with its money-raising efforts and to manage and extract information that would help our government and its intelligence services serve and protect us all globally. In short, to provide a perspective that includes exopolitics in both terrestrial and nonterrestrial environments.)

This proposal divides the $5 million grant to TIFACS and the Monroe Institute. Monroe gets $3 million so they double their operating budget for this year, and TIFACS gets $2 million to begin operations.

The $3 million grant to the Monroe Institute is to enhance the strength of their program and reach. It is also to set their program up at the University of Arizona so that the Gateway Program as designed and offered at the Monroe Institute is offered as a credited course at Arizona.

The $2 million to TIFACS comes with the following structure. TIFACS will be housed at the Monroe Institute and will be managed by the Monroe staff at the direction of the TIFACS staff. They will manage information on the general internet on the Gateway Process on Reddit through the subreddits of "r/astral projection," "r/experiencers," "r/gatewaytapes," and other similar boards where people who have had alien contact can share information. This information needs to be curated at a level beyond the ability of one government and its seventeen intelligence services. This effort in the public sector will provide resources to both the military sector and the science sector and is a force multiplier for our intelligence services. For TIFACS, its five-year mission will require $2 million in the first year and $3 million for the next four years for a total investment of $14 million. Its purpose is to raise half a billion dollars for consciousness research.

The five-year grant to Monroe is to double their budget and reach annually until they can reach 10 percent of the planet's population. So, three million the first year, six million the second year, and twelve million the third year. At this point, we can expect twenty chairs of parapsychology to be funded and a consciousness network of trained psychonauts interacting with the universe. Twenty-four and forty-eight million are projected in the scope of this, but in the fourth and fifth years, this money will come from outside the initial $450 million funding request.

All action taken in this regard is with good intent. This is important when entering an environment where all actors are more powerful than you, with longer histories and better technology. This cannot be a lie, so the policy of TIFACS is one where war on earth must stop. Period.

Higher dimensional intelligences have already indicated they do not want us exporting our hostility into the universe. We still must defend ourselves against those who are not as developed.

The Monroe Institute will use the $3 million they get to set up a training course at the University of Arizona Consciousness Center, and the students there will get academic credit for taking those courses. Additionally, the Monroe Institute will set up a certification program to certify trainers in the disciplines they use.

This training course will be the prototype for courses to be set up at other universities. To begin the twenty chairs of consciousness, we will propose all will have Monroe-type training, and the Monroe Gateway Process will be a credited course at those universities.

Additionally, Monroe will be paid by TIFACS to correlate information obtained by Monroe to better understand the multilevels of dimensional space we are now as a species embarking on.

Other obligations of TIFACS's mandate are as follows:

TIFACS gets a $2 million founder's grant. Its job is to implement the plan to raise $450 million as presented in this proposal. This will be the organization that will manage the overall grant proposal for $450 million. Its mission will be to manage the grants and collect and correlate the data to all organizations that would benefit from said information.

The Scientific Organizational Pyramid: TIFACS

presents funding availability and opportunities for funding to the consortium of schools and organizations through the Consciousness Center at the University of Arizona and additionally through any university that houses a chair of consciousness study or a related field that the study of consciousness would encompass. The choices for programs to be funded are made through and by the scientists themselves in a collaborative manner so that a complete understanding of human consciousness is drawn to the best of our ability now but encompassing all aspects of how our consciousness expresses itself.

The money from the Bigelow endowment to TIFACS gets distributed to the University of Arizona, and they agree to the following terms.

Center for Consciousness Studies (CCS), $10 million

Stuart Hameroff, Center for Consciousness Studies at the University of Arizona.

Stuart Hameroff's research pursued microtubule information processing and anesthetic action. In the mid-1990s, he teamed with eminent British physicist and Nobel laureate Sir Roger Penrose to develop a controversial quantum theory of consciousness ("orchestrated objective reduction" or "Orch OR") based on microtubule quantum computing. Harshly criticized initially, Orch OR is now supported by experimental results, including anesthetic action. In 2017, with Sir Roger Penrose,

The Consciousness Proposal

James Tagg, Ivette Fuentes, and Erik Viirre, Hameroff cofounded the Penrose Institute, intended to support research based on the various works of Sir Roger Penrose (cosmology, quantum mechanics, general relativity, black holes, geometry, and consciousness).

Stuart Hameroff's work is putting the how under our observations and the cognitive landscape we see.

Stuart Hameroff needs more computing power than anyone I know because he is actually trying to count the electrons in the human soul. He would disagree with that rendering, however, but reduced to one sentence, it is what is going on at the Consciousness Center. For your clarity, the electron exists both as a particle and a wave, and everything we talked about in our observations is waveform effects in a particulate environment. Here is the website.[85]

This leads to a new, nonlocal space, where the intelligence in the alien spacecraft we cannot deny any longer resides, just like we do. Billionaires are being asked to fund an acceleration in human evolution. Radin's involvement in DNA tracking of nonlocal consciousness genes foreshadows Commander Troi on a flight deck and humans in space, genetically advanced through clustered regularly interspaced short palindromic repeats (CRISPR).

There is so much more to connect and so much that isn't connected. In the nonmaterialist category, we have

---

[85] https://consciousness.arizona.edu/

stories about interaction with what Brian Weiss calls the "Masters," intelligence forms we can talk to nonlocally. Stuart Hameroff can measure the body during these interactions, and this same holistic approach can be used to ask pointed questions about this science when we do these measurements. So what we are trying to fund is taking Jim Tucker's children, Brian Weiss's regresses, and Pim van Lommel's NDEs into the lab for comprehensive studies. These forms of nonlocal consciousness can be used militarily for data acquisition.

We are coming to a point where our data management capabilities and our bodily information to store all the data of the body will be equal.

Our consciousness of birth and death is an upload-download fractal of information management. This upload-and-download model is a fractal of birth and death, the only difference being the amount of information. This fractal example becomes the methodology for the other fractals.

Monroe/Arizona Partnership

As part of this collective effort, Arizona acknowledges the connection to the Monroe Institute and agrees to set up a program with Monroe to jointly proceed with tests and experiments as may be designed and required by the joint Monroe/Arizona program. Each commits $1 million of the funds of this initial grant to this effort.

The University of Arizona College of Social and Behavioral Science ($5 Million)

The College of Social and Behavioral Science, where the CCS is housed, has to be funded to study the global effects of our coming change of consciousness in terms of religion, society, and politics, as completely as we can get it, to publish *The Standard Model of Consciousness,* then teach it.

I want to make clear to you the importance of this particular contribution/donation and the point of keeping it separate and distinct from the contribution to the Consciousness Center. The school is/will be the central processing point for the information from the Consciousness Center. It is here that the information will be taught, and the impact of the new information about consciousness will be evaluated, implemented, monetized, and weaponized. It is a central clearinghouse for how the basic science is done at the Consciousness Center; it standardizes how the science is processed, refined, and used. You can instruct the religious on how to rewrite the human belief system from this point because it will turn belief into understanding.

What follows is a simple factual account of what you could call the standard model of consciousness, which all the contributors collectively produce. It posits that we live multiple lifetimes, our consciousness resides in and outside our bodies, and when we die, our consciousness is still discreet, meaning it is still "you." The math proof changes religion because we have a better scientific understanding of the afterlife. However, religious law must be rewritten in many cases to accommodate our new factual knowledge. It changes our view on sexuality because life to life, we change gender regularly. As we incorporate this new view of consciousness into our reality, we encounter aliens who fly through our airspace, which we have documented. The Department of Defense is looking for new methodologies that will put them into communication with those aliens on new wavelengths. Your donation and the donations of the other donors will make all of this possible.

The Asks

I would ask the College of Social and Behavioral Science

The first ask is for them to compile and complete a book titled *The Standard Model of Consciousness* in preparation for writing its curriculum. This needs to be taught at all levels. In it, they have to present an inclusive look at our reality, including, and not limited to, the points in this proposal—such as children who remember previous lives, past-life regression, NDEs, remote viewing and clairvoyance, and all other understandings as best presented at the Consciousness Center. This includes contributions by the scientists named in this document.

The second ask is to evaluate the social and political

consequences of a reality that includes aliens and an afterlife. The ramifications must be part of the book *The Standard Model of Consciousness.*

The third ask is to develop a plan to deal with world political change because of this new reality. That includes repairing the damage to the earth and seeking funding to enact that change.

The fourth ask is to teach this information at every level, globally. This involves Prometheus Entertainment, who already have the footage and the understanding of this reality. Once done, it can be translated into every language, and people can watch it on their cell phones. I am asking billionaires to fund and implement global cognitive change for the sake of humanity.

The ask of Prometheus Entertainment would be paid for by the College of Social and Behavioral Science, which would be writing the curriculum for this new iteration of consciousness science. The interaction between College of Social and Behavioral Science and Prometheus would begin with a video prototype for classroom presentation written jointly by the College of Social and Behavioral Science and Prometheus Entertainment to present all the information on the science of extraterrestrials and aliens and the various subcategories of each.

I would ask them to work with Prometheus Entertainment to produce educational videos on the standard model of consciousness and its relevance in society for every variation of consciousness and in every language. In putting it on video, it can and will end up on every cell phone on the planet.

For the first time, with a billionaire's contribution, there is cohesion among the disciplines of consciousness science, and the document produced becomes a teaching tool to a new generation so arguments advanced by J. K. Rowling and others can be refuted scientifically and mathematically. Dissemination of this information needs to go to the harshest environments for LGBTQI+ people. Therefore, we ask that this be printed in all languages and have a home on the Web at www.tifacs.org. We ask the following organizations to disseminate copies of these produced documents in areas where they are trying to save individuals so that, ultimately, the mindset changes globally.

At that point the commanding general of the Army Futures Command asks his subcontractors, who have produced no harvestable weapon system for $32 billion, to send TIFACS $100 million.

This finishes the first tranche and the second $50 million build-out. That build-out is planned and executed by the best and the brightest in the consciousness field, the authors who responded to the Bigelow essay contest, the judges who judged that contest, and the University of Arizona Consciousness Center, following Radin's plan.

What follows then is the rest of the $30 million from the first tranche.

* * *

The Consciousness Proposal

The science of Luke Ruehlman's story in chapter 4 tells us that the human life force or soul has no gender, social, or racial identity. The odds-against-chance calculations suggest that Pamela Robinson returned as Luke Ruehlman. This is not an unusual case study.

Hanan Monsour died and came back as Suzanne Ghanem and started, at sixteen months, recalling her old life. When reunited with her prior family as Hanan, Suzanne identified fifteen prior relatives by name by the age of five. Those family members came to accept Suzanne as Hanan.

Charles Leininger, as a child, remembered dying as a navy pilot during World War II in the Pacific. He remembered the ship he was on, the *Natoma*. He was James Huston in that life, and when young Charles was taken to a reunion of *Natoma* veterans, they accepted him as who he was in his previous life.

I could go on with more examples, but they all amount to the same thing: The odds against the chance of these people knowing about their previous lives is certain. Being able to convince others of who they were occurs because they know things private to the person they are trying to convince. They *were* who they claimed to be; both parties remembered the same facts in the same way because they shared a common experience.

These stories are examples of a category of consciousness science called children who remember prior lives

(CWRPL). The UVA has been studying cases like this for over sixty years.

This kind of story was once considered a scientific anomaly, something our science could not explain. But once they are connected to other categories as fractals, such as NDEs and past-life regression, they prove mathematically that this is our reality.

I am asking you for several smaller interconnected donations whose benefit will add up to more than the sum of their parts. Once complete, its use will be global. The following explanation will also show where we are asking others to support this effort. All funding requests, including this one, will be available on the website of TIFACS. The plan will also be there, laying out the need for $450 million for consciousness research.

The overall need request is $450 million and was written by Dean Radin.[86]

This plan emphasizes four areas: education, empiricism, theory development, and applied research.

This request to you, each billionaire, is for $20 million. What follows is how and where that $20 million will be spent and on what. Each recipient is a luminary in the field of consciousness science, each with a different focus. Our mission at TIFACS is to infuse consciousness science with resources and build connectivity to increase and improve

---

[86] Bob Good, Dean Radin, Stephan A. Schwartz, Titus Rivas, and Cathie Hill, *The Mathematics of the Science of Reincarnation* (Boynton Beach, FL: IASOR, 2020), 373–385.

our understanding of ourselves and each other. *I would trust each recipient would do as this plan requests or return the gift/donation.* Each donation is given to the named individual to manage for the benefit of consciousness science. They all know of this effort to a greater or lesser degree.

* * *

Request for Funding Jim Tucker, the University of Virginia, Children Who Remember Prior Lives, $5 Million

Jim B. Tucker, MD, Director of the Division of Perceptual Studies, University of Virginia Health System. He studies children who remember prior lives.

Two things: First, when we use odds-against-chance calculations through thousands of cases, the fact that our consciousness transcends our permanent bodily death is now a calculable certainty in each category of this science. When we use each category as a fractal, we get mathematical proof that the story you just read is true; we proved it and a new reality with it. With that mathematical proof, we make the case that souls can change gender, race, and religion from life to life. That scientifically makes the case that LGBTQI+ is normative. That—scientifically—apostasy, blasphemy, and celibacy should be removed from all religious canon. Scientifically we can use this information to reshape our society.

The Applications of the Science of Reincarnation

The request is for $5 million to support this research at UVA, and the ask is for Jim to take out, in one compendium, the transgender cases he and the late Ian Stevenson have compiled, like the first example. Using odds-against-chance calculation, these observations are a certainty. Still, currently, there is no way to connect them to our reality scientifically, even though it could be done, and prove scientifically that being gay, straight, or transgender is just a common experience of consciousness and not a disorder. Their observations alone can't prove that, but if he aggregates them, we can connect them to other observations in consciousness science and make that case.

So, we are providing a list of requests that your contribution will thread together to present this science in a cohesive and easy-to-understand explanation so that all causes you support can use this information to refute the lies and lack of knowledge on the subject and to educate the world.

The asks: We are asking Jim Tucker to lift and separate:

1. Transgender narratives
2. Narratives by region (Mideast, Far East, West, etc.)
3. Narratives by religion
   This is to show that the effects we are documenting are universal.
4. These will be contributed to the book *The Standard Model of Consciousness*, being cataloged by the

University of Arizona under the observations section, and their mathematical information will be given to Jessica Utts so she can create and produce a fractal proof.

To that end, we want to aggregate the landscape of observations and lift transgender stories from the data pool/mine because this is our reality proven by mathematics.

* * *

Brian Weiss, the University of Miami, Past-Life Regression, $5 Million

This category is past-life regression. Rather than present a case study here, refer to the past-life regression of actor Glenn Ford as our example. Simply google it.

Brian Weiss graduated from Columbia University and Yale Medical School and is chairman emeritus of psychiatry at the Mount Sinai Medical Center in Miami.

The gift would be to the Mount Sinai Medical Center Psychiatry Department at the University of Miami.

Brian Weiss, $5 million, and the ask is to do the same thing we are asking of Jim Tucker. He regularly encounters the lives of the opposite gender when regressing individuals. Ultimately, all these are observations of how consciousness manifests itself, and eventually, they become data points in our fractal proof.

The asks: We are asking Brian Weiss to lift and separate:

1. Transgender narratives
2. Narratives by region (Mideast, Far East, Western, etc.)
3. Narratives by religion
   This is to show that the effects we are documenting are universal.
4. These be contributed to the book *The Standard Model of Consciousness*, being cataloged by the University of Arizona under the observations section. Their mathematical information will be given to Jessica Utts so she can create and produce a fractal proof.
5. Ask questions of the "Masters" regarding nonlocal consciousness and suggestions for studying it.

* * *

Pim van Lommel, Category: Near-Death Experiences, $5 Million

NDEs are cognitive events that occur when a person experiences bodily death. While there is not a transgender narrative here that I know of, whether you are gay or straight, you can experience an NDE. The landscape that the person experiencing the NDE describes is very similar to the landscape that CWRPL and past-life regression describe. There are enough data points between these groups to identify the landscape described as fractals of each other.

Explaining NDEs, this is what Pim van Lommel says about the continuity of our consciousness:

> According to our concept, grounded on the reported aspects of consciousness experienced during cardiac arrest, we can conclude that our consciousness could be based on fields of information consisting of waves and that it originates in the phase-space. During cardiac arrest, the functioning of the brain and other cells in our body stops because of anoxia. The electromagnetic fields of our neurons and other cells disappear, and the possibility of resonance, the interface between consciousness and the physical body, is interrupted.
>
> Such understanding fundamentally changes one's opinion about death, because of the almost unavoidable conclusion that at the time of physical death consciousness will continue to be experienced in another dimension, in an invisible and immaterial world, the phase-space, in which all past and present and future is enclosed. Research on NDE cannot give us irrefutable scientific proof of this conclusion, because people with an NDE did not quite die, but they all were very, very close to death, without a functioning brain.

The conclusion that consciousness can be experienced independently of brain function might well induce a huge change in the scientific paradigm in Western medicine and could have practical implications in actual medical and ethical problems such as the care for comatose or dying patients, euthanasia, abortion, and the removal of organs for transplantation from somebody in the dying process with a beating heart in a warm body but a diagnosis of brain death.

There are still more questions than answers, but, based on the theoretical aspects of the obviously experienced continuity of our consciousness, we finally should consider the possibility that death, like birth, may well be a mere passing from one state of consciousness to another.[87]

This theory, as articulated by Pim, is supported by the narratives above. They are, in fact, fractals of each other. They are each an iteration, and they are self-similar—the donation to Utrecht University should be balanced by the following asks.

---

[87] Pim van Lommel, "About the Continuity of our Consciousness," in *Brain Death and Disorders of Consciousness*, ed. Calixto Machado and D. Alan Shewmon, vol. 550, *Advances in Experimental Medicine and Biology* (Boston: Springer, 2004), 18, https://www.rcpsych.ac.uk/docs/default-source/members/sigs/spirituality-spsig/pimvanlommel_about.pdf?sfvrsn=cb878f8c_4.

The asks:

1. Transgender narratives
2. Narratives by region (Mideast, Far East, Western, etc.)
3. Narratives by religion
   This is to show that the effects we are documenting are universal.
4. These be contributed to the book *The Standard Model of Consciousness*, being cataloged by the University of Arizona under the observations section. Their mathematical information will be given to Jessica Utts so she can create and produce a fractal proof.

We ask Pim to provide Jessica with matching fractal points from his observations that match the other categories. NDEs happen to transgender people, in all races and all religions. The effects we are documenting are universal. These are contributed to the book *The Standard Model of Consciousness*, being cataloged by the University of Arizona under the observations section.

When we take all three views of this nonlocal consciousness, we get different perspectives of the same site, looking forward (remote viewing) and backward (past-life regression). We also see that same landscape in the NDE observations.

How do the mechanics of this landscape and our travel through it work? Does this go directly to communicating with aliens in our local space?

There is one final notation to this: From the sum gifted to Utrecht University, 10 percent of that sum of $500,000 be set aside for Titus Rivas to use at his discretion in the advancement of out-of-body consciousness.

* * *

Jessica Utts, Chair of Statistics at the University of California–Irvine, to formalize the Math, $5 Million

She has done odds-against-chance calculations of the various subsets we have discussed, but while that math points to certainty, it is not a proof. If she does the fractal geometry of her calculations, she "proves" this hypothesis mathematically.

Now several exciting things happen with this proof. First, it challenges any other evidence to be more factual and supports religious belief in an afterlife. This condition of science also includes all creatures with electromagnetic signatures, so all religions are present; it eliminates apostasy, blasphemy, and chastity as anything other than wrong and not founded in fact.

The donation would be to the University of California at Irvine.

The asks: We ask Jessica Utts the following:

1. Provide the framework for the fractal proof that this is our reality. The landscape of observations will be aggregated and shown to be valid across all races and cultures. Transgender narratives, narratives by region (Mideast, Far East, Western, etc.), and narratives by religion we are documenting are universal.
2. These will be contributed to the book *The Standard Model of Consciousness*, being cataloged by the University of Arizona under the observations section.
3. When we take all three views of this nonlocal consciousness, we get different perspectives of the same site, looking forward (remote viewing) and backward (past-life regression).

* * *

Noetic Institute, $5 Million

Let us say that there was a group of scientists who wondered if there was a DNA marker that indicated if someone was "gifted" when it came to nonlocal consciousness, and suppose that these scientists had DNA samples taken from the best psychics they could find. It would be cheap to take the sample and send it to a lab. Let us also say that these scientists identified the DNA code for such an ability, much like finding the DNA marker for sickle cell anemia. And let us say that using CRISPR could enhance that nonlocal ability. In a simple experiment, they would be rendering *Homo sapiens* into the past and introducing *Homo futurists.*

The asks: I would ask Radin and the Noetic Institute to contribute to *The Standard Model of Consciousness,* and the $5 million is to further their work.

That's the first tranche to fund Radin's plan for $450 million. Here is the link to the overall plan:

http://www.thescienceofreincarnation.org/wp-content/uploads/2020/11/DeansPlan2.pdf

***

The World Professional Association for Transgender Health (WPATH), $5 Million

The World Professional Association for Transgender Health (WPATH) promotes the highest standards of health care for individuals through articulating standards of care (SOC) for the health of transsexual, transgender, and gender nonconforming people. The SOC is based on the best available science and expert professional consensus.

The overall goal of the SOC is to provide clinical guidance

for health professionals to assist transsexual, transgender, and gender nonconforming people with safe and effective pathways to achieving lasting personal comfort with their gendered selves in order to maximize their overall health, psychological well-being, and self-fulfillment.

The ask: WPATH should use the literature we produce through TIFACS to educate in areas that persecute people of gender variance.

* * *

Rainbow Railroad, $5 Million

Rainbow Railroad is a global not-for-profit organization that helps LGBTQI+ people face persecution based on their sexual orientation, gender identity, and sex characteristics. In a time when there are more displaced people than ever, LGBTQI+ people are uniquely vulnerable due to systemic, state-enabled homophobia and transphobia. These factors either displace them in their own country or prevent them from escaping harm.

Instead of just getting people to safety, they distribute to those points in the world the aggregate of the information included in this fund request.

The ask: They distribute *The Standard Model of Consciousness* to explain that gender variation is normative among humans and not a disease; it's a common manifestation of consciousness.

* * *

Who is being asked and why?

All these efforts must be coordinated to produce *The Standard Model of Consciousness* in textbook form, teach it, and advance the solutions to human management it prescribes. There are several questions we need to address before we begin this.

1. If this is so important, why aren't we advancing an intelligent plan and facing squarely the reality unfolding before us? No political system on earth can address this.
2. Of the individuals who can see aspects of this consciousness research in all its forms, how are the efforts being coordinated?
3. Of the emerging paradigm, what is the response from the best minds we have to deal with this?
4. The point of everything I have written is to coordinate this effort of self-defense and self-discovery for a species: humanity.

Who would I initially invite to a consortium of billionaires?

Laurance Rockefeller. Billionaire Laurance Rockefeller is financing what is billed as a scientific survey aimed to sort hoaxes from the purportedly truly mysterious crop circles.

Since 1978, patterns of matted-down cornstalks, rapeseed, and other crops have been appearing in the fields of Wiltshire and neighboring areas in England. Enthusiasts attributed the so-called crop circles to UFO landings. Many people still believe that some of the patterns, because of the speed of their appearance and their perfect geometry, must have a nonhuman origin

There are two distinct varieties of crop circles and how they are created: one a human version of hucksters and the other ones made by means we do not understand, with an obvious signature of which is which. But what information is being exchanged?

Robert Bigelow. Robert Bigelow has a deep understanding of this landscape and knows how to manage this type of venture within consciousness science. Begin the plan to fund twenty chairs of parapsychology using your judges from the writing contest you just ran and the Center for Consciousness to select where the chairs should be placed and what funding they should get. I would ask that the first chair be placed at the University of Edinburgh and be called the Dumbledore Chair. If each chair is funded for $5 million, this would need

$100 million. See Radin's plan for information on the structure.

Melinda Gates. I would ask to become involved with education as outlined in Radin's plan. The science we are defining rewrites human history and begins to explain our place in the universe. Someone needs to take ownership of the global distribution of this information, and no one is better qualified than Melinda in this endeavor.

Mackenzie Scott, your $20 million donation toward the $450 million we need to fund consciousness research goes directly to education. Radin's plan is attached here. Your donation funds TIFACS to run other programs and other projects to seek those contributions to fill out the list of Radin's proposed needs.

Bill Gates. There is none better to lead a global information processing effort, including AI and nonlocal consciousness. Please see chapter 14, "The Remote Artificial Intelligence Viewer" in the book *The Mathematics of the Science of Reincarnation*. It will open the universe to communication, like how the radio opened the earth to instant communication. You should sit on the applied technology board and the theory development board.

Jeff Bezos. This proposal is nothing less than creating a standard model of consciousness so that we can protect ourselves collectively from the threats we all face and some of the risks we will have to take. In that, there is scientific research demonstrating that element 115 reacts to gravity. Imagine creating a motor that does not need fuel, rides on gravity waves, and has endless, ever-present energy in abundance. One technology change to a gravity drive changes everything. Can we access nonlocal sources for the how-to? Once done, where in the universe will it take us? You should sit on both the applied technology board and the theory development board.

Mark Cuban. In a sentence I want you to commercialize anything that falls out of the research and application that comes from the consortium's work. I also want you to sit in the applied research chair.

By collaborating, we can contribute $450 million and manage the threats and opportunities of this new reality.

Mark Cuban, you would become involved with the Consciousness Center to fund commercial applications of discoveries.

I mean no disrespect when I attach this UN report that says we are headed to an unlivable world. Consciousness research is critical to the solution. Your money will not protect you from a collapsing world, but it might save us all.[88]

---

[88] Frank Jordans and Seth Borenstein, "UN Warns Earth 'Firmly on Track toward an Unlivable World'," AP, published April 4, 2022, https://apnews.com/article/climate-united-nations-paris-europe-berlin-802ae4475c9047fb6d82ac88b37a690e.

Charles Koch. You are being asked to help rewrite the Russian Constitution in the next chapter to align it with Western constitutions, a place where leaders retire from public life rather than being overthrown or dying in office. This is not to overthrow the Russian government but to leave the public officials in place and institute a freer model to improve commerce and health. The goal is to provide a level of benefit to their population compared to Finland. While the initial ask of you is to help rewrite the Russian Constitution, the underlying is to help design a demountable constitution that will harmonize the world's constitutions to facilitate trade and the safety of economies. This is fully explained in chapter 16.

Its basis should be in an operating system design; in fact, I would like a version designed by systems engineers rather than politicians, as all 195 countries need to be harmonized if we are to "Face Space" together. The United States Constitution must be an example of removing gerrymandering so true sample size components can optimally and frictionlessly operate in an interplanetary environment. This is not to remove autonomy from individual groups but to present the best operating system that they can adopt, plank by political, social, and religious plank. The consciousness science which will be embedded in the constitution I am asking you to fund will have the fluidity to facilitate major social change without political upheaval.

The Consciousness Proposal

Finally, to the bankers who might read this. The overall proposal presented here makes the case that consciousness research is a cottage industry. This proposal is a roll-up in a banking business sense, and the whole of the result will be greater than the sum of its parts. The value of these resources will be exponentially greater after the infusion of this funding, the ramifications of which will be that humanity is off-world and dealing with other civilizations. People become an export, and our global GNP will be a measure of our racial worth by other races. Great wealth will simply be lines of code. War here must stop because it represents poor social management. Those resources need to be redeployed. This plan needs another $450 million. My part of the plan is getting Radin his $450 million with no responsibility for ROI and the next $450 million to deal with the ramifications.

I am asking for $1 billion, so I want to make this simple for you. I want to reduce the opportunity to two simple ideas so you will buy in—I don't want a contribution; I want buy in: Aliens communicate in a common telepathic language. Something humans, if trained, can do. In chapter 10, "Education," I will explain the need to found the school for alien studies because I will discuss how they are communicating through crop circles that cannot be hidden, and the images are those of chemical compounds like niacin depicted in the crop circle on August 1, 2013, at Devizes, Wiltshire. Another is the lithium clock of August 13, 2018.

Decoding these ongoing messages reveals that the chemical images they are sending have very significant meaning to our collective safety. The channel is open and the communication real. They are transmitting better technology through these images. And collectively these images paint a picture of our future. There is money in funding these developments but that's not why you need, yes, an imperative, to invest.

The tobacco companies sold us tobacco knowing full well it caused cancer. You can't rely on a few men who are dealing with aliens for all of us to continually make decisions in our best interest. This threat affects every dime you have, and the collective consortium of global scientists is the best way you have both to make money and protect your collective asses.

This is what TIFACS will attempt to do while raising the $450 million for consciousness science. Here is the link to other initiatives that TIFACS will make in the coming months: www.tifacs.org

It should also manage both the website and YouTube channel. They currently look like this: www.tifacs.org. It is where the framework for launching this effort is being built. Your contribution formalizes this effort. You can connect to the YouTube channel there or the Reddit subreddit "r/reincarnationscience." There is a list of different domains pointing to this site, and it is where the science we are asking to be aggregated should be given to all interested organizations for free, supporting various causes connected to this effort.

This will:

- Provide a fact-based narrative to the world in print and film.
- Fund consciousness research.
- Deal with the aliens in our airspace in a cohesive and managed way, something governments are incapable of.
- Deal with the continuity of consciousness; if we are to return to Earth in another life, may the next one be free of war. This is the goal.

This heartfelt and reasoned cause needs resources; the resource I am asking for is you. I want you billionaires to be its guardian angel.

This will change the world. We ask that you share this request with the other donors so we can collectively work together. This document also has asks of the scientists and organizations named here. We aim for total transparency in how we operate, what is funded, what the results say, and how we make decisions on further development based on results gathered to date. We need the brightest minds, and we need resources to have them incentivized to work with us.

This is a team effort of donors and researchers to

produce a standard model of consciousness, how to use it, and the consequences and ramifications of such an effort.

All donations go directly to the institutions named above.

Homer, TIFACS is the operational center for organizing and distributing money collected for this project. A new class of soldiers will emerge from these efforts, and the group will be known as the Clairvoyant Space Corp, and its individuals will be known as psychonauts. It will be larger and better funded than anything we have now. Want to buy a mug?

Homer asked who Bob is. It could be Bob Monroe, and it could be Bob Bigelow, and it could be Bob Good. It is all of them, and all of you.

Chapter 10

Education

The next four chapters must be taken as a set rather than separately because they are so interconnected. I will address the billionaires, and Dean Radin will address the scientists. We are both saying the same thing, but it must be explained differently to each group.

In this chapter, I will prove—yes, prove—that aliens are communicating with us directly, explain the immense disinformation program being run, and explain why this information block must be dismantled for everyone's benefit—ours, our government's, and interestingly, the aliens' benefit as well. It must be done not in opposition to our government but to support it and bring new resources to our global problems.

This is my request for Mark Cuban to sit on the applied technology board at TIFACS; for Melinda Gates and Mackenzie Scott, along with Robert Bigelow, to sit on the education board; and how to begin, and that Robert Bigelow, once the process has begun, to move to the theory development board.

This also should explain why aliens use crop circles to communicate with humans.

I have already explained crop circles and the Grays' history with the American government. We all read and listen to the news, and the government refuses any information on disclosure while whistleblowers come out with bits of information that are then discredited or denied.

So let us go back to Frank Drake and Carl Sagan. They sent the Arecibo message in 1974. In 2001 we received the Arecibo answer as described here by Wikipedia.

This is Wikipedia's information on the Arecibo answer on their page in 2024, twenty-three years after the appearance of the Arecibo answer.

> The "Arecibo answer" is a crop circle (well, crop rectangle, to be accurate) that is purported to be a response to the "Arecibo message," a piece of coded information about Earth and humanity which was first beamed into space in 1974. It appeared in 2001 near the Chilbolton radio telescope in Hampshire, UK. Despite the fact crop circles are *known* to be hoaxes, people still believe that this is key evidence of extraterrestrial presence on Earth. At the time of this hoax, the usual suspect came out of the woodworks to claim that this is "undeniable proof!!!!" that aliens have contacted us. Amidst all this madness, SETI had to actually come out and spell it out that the obvious hoax is indeed an obvious hoax.

Education

As I went through the rewrites of this book, I needed to go back to Wikipedia to footnote what they said and found the entry had changed. What they say now, which I will footnote, is the following.

> "The SETI Institute Online rebutted the idea that this was a genuine extraterrestrial response, by saying, "This is highly improbable. There is no evidence to suggest an other-than-earthly origin for these graphics."[89]

They also list crop circles as hoaxes as evidenced by the web address in the footnote.

I want to address the false narrative being presented here as it directly affects those who might fund this research, and the Wikipedia entry does not explain the successive messages we are receiving or their interpretations.

While there is no exact count, since 2001, thousands of crop circles have appeared worldwide, with the majority occurring in the UK, particularly in Wiltshire, where the phenomenon is most prevalent; reports suggest that hundreds, if not thousands, of new crop circles are created each year, with many believing most are human-made rather than paranormal in origin.

---

[89] "Arecibo Answer Crop Circle Hoax," Wikipedia, accessed January 30, 2025, https://en.wikipedia.org/wiki/Arecibo_message#Arecibo_Answer_crop_circle_hoax.

Since 2001 crop circles have become more intricate in design, featuring complex patterns and multiple shapes.

While Wikipedia acknowledges the increased complexity of crop circles, it misattributes the analysis, to show how crops are raised in circles to irrigation and so forth.

Let me explain the situation. The Grays want to be known to Earth. The American government made this information top secret and refused to disclose its relationships and treaties with alien races. Hence the very powerful cover-up, down to controlling information on Wikipedia and elsewhere.

The reasons for resistance within the government to disclose this include national security, global destabilization, and a host of other fear-pandering reasons to not disclose. In the meantime, technology leaps forward, enriching certain factions.

Some alien groups fear for us and our well-being. Earth is a way station in space. Aliens have lived here long before us and still do. Some care not for humans at all and to some humans are a long-term resource. The Grays fell into that category.

How do the Grays announce their presence to all of us to circumvent our own government obstruction? How do they communicate with everyone to show their non-aggressive intent? So, crop circles are billboards that last for several years in that the shadow of the crop circle is still visible in the ground in succeeding growing seasons.

They have chosen to put up these billboards that cannot be obstructed. They also do not want global upheaval. They have time on their side; they have been here for thousands of years and watched us develop.

They are concerned for our future, not for altruistic reasons but the loss of a major asset that is being threatened. By whom, you ask? The answer is us. By our actions, we are destroying our planet, and global warming threatens our species by reducing our fertility. How can these threats be addressed for the benefit of us all?

Here is a fair question for our intelligence services: Why have we let our radio telescope at Arecibo fall to ruin while the Chinese have erected one twice the size of the one at Arecibo? The bandwidth of how many Earth organizations are communicating with extraterrestrials is expanding, and the organization proposed here brings more resources to your efforts. This asks for no disclosure; this proposal asks for intelligent analysis and response, funded by our richest, to fund our best.

To be clear, not every alien group has good intent, but exopolitics is very complex with nonbiological entities having input in humanity's future.

Why, in a book titled *The Applications of the Science of Reincarnation*, would we be talking about nonbiological entities when we are discussing extraterrestrial politics? Because the science of reincarnation maps fourth-dimensional consciousness, and that telepathic language connects all of us.

This is why the School for Alien Studies must be founded at the University of Arizona, where the following work will begin.

Mark Cuban, this part of the proposal is directly to you because of your involvement in reducing drug prices for everyone's benefit. I am asking you to expand that by funding this research to develop a new drug to counter the damage we are doing to ourselves, and I want you to follow the pathway to do this as explained in Jerry Kroth's book *Messages from the Gods: A Scientific Exposition on the Extraterrestrial Origin of Crop Circles.*[90]

Mark, for all the disinformation, crop circles are of alien construction, and here is why. They have begun to announce things no one on Earth could have known about. Two crop circles appeared on July 13, 2011, one on Chaddenwick Hill and one on Avebury Trusloe.

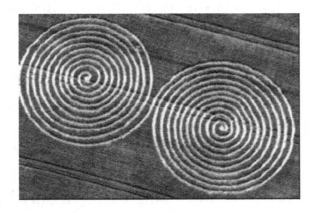

---

[90] Jerry Kroth, *Messages from the Gods: A Scientific Exposition on the Extraterrestrial Origin of Crop Circles* (Genotype, 2022).

Education

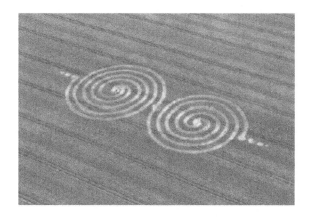

For a full explanation see page 204 of Messages from the Gods. There is information coded in how the spiral lines flow into one another, both clockwise and counter-clockwise.[91] The point here is the information contained in these images was of a magnetar, a type of neutron star that gives off gamma-ray bursts. The crop circles appeared showing such a gamma-ray burst. No one on Earth could have known this would be detected the next day. Kroth's question is similar in some ways to Moody's. "If the magnetar was observed here on earth on July 14th, how could one have known about it 24 hours earlier?"[92]

This is disclosure, but not from our governments. This is disclosure from the aliens themselves.

That is just the opener because some crop circles depict cosmological events. The two crop circles—one on

---

[91] Jerry Kroth, *Messages from the Gods: A Scientific Exposition on the Extraterrestrial Origin of Crop Circles* (Genotype, 2022), 205.

[92] Jerry Kroth, *Messages from the Gods: A Scientific Exposition on the Extraterrestrial Origin of Crop Circles* (Genotype, 2022), 218.

Martinsell Hill on June 25, 2009, and one on Wayland's Smithy, Oxfordshire, on May 29, 2009—depict a collision. On July 19, 2009, amateur astronomer Anthony Welsey was the first one to see the comets that were about to impact Jupiter. Again, information ahead of the event.

This is not why I am asking you to sit on the Applied Technology Committee. It is because of the chemicals the aliens—I believe are the Grays—are sending us. Crop circles depicting astatine, polonium, cesium, lithium, niacin, and so forth, and Kroth has related that to a decline in human sperm production. I will not go into the global warming aspect of the explanation but to say these images are designed to help us, and if the planet goes to hell, then every billionaire I am asking to help goes along for the ride.

The crop circle on Roundway Hill on July 23, 2011, depicts the chemical structure of melatonin.[93] Kroth thought, "Human-induced alterations in your planet's climate will have a negative effect on your progeny."[94]

The crop circle on May 29, 2020, was a rendition of the COVID-19 virus that depicted things discovered later.[95]

---

[93] Jerry Kroth, *Messages from the Gods: A Scientific Exposition on the Extraterrestrial Origin of Crop Circles* (Genotype, 2022), 50.

[94] Jerry Kroth, *Messages from the Gods: A Scientific Exposition on the Extraterrestrial Origin of Crop Circles* (Genotype, 2022), 63.

[95] Jerry Kroth, *Messages from the Gods: A Scientific Exposition on the Extraterrestrial Origin of Crop Circles* (Genotype, 2022), 101.

Education

The image is taken from the Crop Circle Connector,⁹⁶ which is a valuable media site for this type of information. As you read this, Mark, I want you to see the curriculum for the school of extraterrestrial studies.

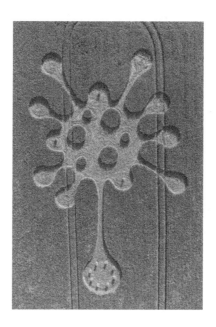

There are nine protrusions, but I want you to look at the smaller pegs inside those protrusions; they are called dimmers.⁹⁷ Kroth says, "The most important feature of this crop circle is that it performs an amazing, magical feat. It enlarges a single spike protein (the bottom blob) and gives us a chance to take a good look inside...and here we find 8 perfectly drawn pegs."

---

[96] https://www.cropcircleconnector.com/nonflash.html
[97] Jerry Kroth, *Messages from the Gods: A Scientific Exposition on the Extraterrestrial Origin of Crop Circles* (Genotype, 2022), 90–123.

The Applications of the Science of Reincarnation

Scientists call that 'Sp8'. And that, in turn, led us to scientific literature that said Sp8 had an immunological function, that it might 'elicit neutralizing antibodies in Covid 19 patients.'

And that is exciting news. That is a clue. That is a revelation."[98]

"This crop circle is urging us on. It is one step ahead of our research. It is pointing us in an important direction."[99]

Now realize this information is intended to help us and think of the possible benefits to mankind if instead of communicating through crop circles, we opened a nongovernmental way of communicating directly to accept help for our species from aliens who are displaying the intention to help. This is disclosure but not acknowledgment from our government.

At the end of his book, Kroth presents a simple proposal: It needs to be funded.

"Imagine a government-sponsored grant. There are perhaps 80 crop circles that appear every year."[100]

Step 1: So, imagine teams of three persons each, ten of

---

[98] Jerry Kroth, *Messages from the Gods: A Scientific Exposition on the Extraterrestrial Origin of Crop Circles* (Genotype, 2022), 113.

[99] Jerry Kroth, *Messages from the Gods: A Scientific Exposition on the Extraterrestrial Origin of Crop Circles* (Genotype, 2022), 114.

[100] Jerry Kroth, *Messages from the Gods: A Scientific Exposition on the Extraterrestrial Origin of Crop Circles* (Genotype, 2022), 515.

them, were on these assignments and each team had a budget of $15,000 to immediately fly out to the reported circle and carry out its analysis; the cost would be $1,200,000.

Step 2: A cadre of scientists is assembled: five physicists, five chemists, five microbiologists, five mathematicians, topologists, and geometricians, five linguists, and/or specialists in hermeneutics, and five social scientists. Each would receive $100,000 in a single year for half-time commitment to be involved in the crop circle project. In addition, they would each have a budget of $10,000 to help them with their own analysis.

Step 3: An independent review board would analyze their submissions.

After everything is added in and doubling the money for a two-year commitment, the total cost estimate is $10,600,000.

So let us talk about why, where, and how to spend another $100 million to organize this. The structure must be built out and clear.

This money will found the School of Extraterrestrial Studies and begin the conversation with another species by decoding and responding to the information in the exchange that has been going on for years.

The following is to Mark Cuban:

Mark,
In a sentence, I want you to contribute $20 million to this effort and commercialize anything that falls

out of the research and application that comes out of the consortium's work. What we are proposing here is a research lab on a global scale with transcendent results.

At its core, it gives Hameroff and Bigelow fifty grants to the grassroots of this science. They decide where this goes. Just creating the twenty chairs to this science alone will be game-changing.

This letter to you is part of a series of letters and initiatives designed to do just that, fund consciousness research.

At the same time, there has been and will be continued alien contact and information passed between alien species and our own. We need to be cognizant of that as a species and the blocks to that information being passed removed. Personally, Mr. Bigelow, that is at the core of your search, and I am not just asking you for $20 million; I am asking you to direct the spending of $500 million in that search. Setting up the twenty chairs of consciousness is simply improving the network of research needed to fulfill the CIA's and the alien's desires, connecting our minds to the machines we pilot through the universe. During that, our navigational ability in the dimensional manifold will cross the intersection of your past that you seek and will be there when you need to visit. While I can guarantee it is there, it is only reachable in its dimension, and that is ethereal.

Conjoining what the religions want you to believe and what the technology and emerging model of consciousness would support is something we can no longer ignore. It is awareness in another dimension.

Through this dimensional manifold, we have a window into cognition in a different waveform in time-space. The ramifications change how humanity should manage itself and our world because this dimensional manifold gives us scientific access to a region of time-space that we could only believe existed. This information impacts data processing, space travel, and human development and design through AI (artificial intelligence) and clustered regularly interspaced short palindromic repeats (CRISPR), which form the basis for CRISPR-Cas9 genome editing technology and go directly to human governance and development.

This cognitive science goes to national defense because it exposes cognitive movement in a dimensional manifold. In fact, space travel and the afterlife are joined scientifically through cognitive intent and a dimensional intersection between the third and fourth dimensions.

By time having mass (mass being defined as ordered information), you have objectively started mapping a fourth dimension we can see into. In doing so, you laid the groundwork for a propulsion system through time and navigation of that area of time-space. Don't you think we will encounter cognitive life in a fourth, higher dimension when it exists here in profusion in the other three? If you distill the exit interviews of alien abductees, nearly all were communicated with telepathically. That communication is nonlocal, and the odds-against-chance calculations are the same as Moody's on NDEs. Certainty.

This science is evolving faster than we are keeping up. With one leap, it can be weaponized against us at any point from any point. These are common threats to us all. Our effort in response must be collective, cohesive, and immediate. This is a call to academic arms, and the only way we can do this research efficiently is to do it together. *The Applications of the Science of Reincarnation* presents a malleable and inclusive plan from which we can all participate and benefit.

So, Mark, what is the commercial value of owning the rights to a reincarnation model of information transfer under a new hypothetical company called Bio Data Services? We have the math proof; we see it being used by extraterrestrials, the Grays; and we cannot have this new science model without changing our views of consciousness. I ask you to call Stuart Hameroff and give him $5 million to be given out in twenty grants, so each school with a chair gets $250,000 to begin this cohesive and comprehensive program.

So, Mark, here is my formal request to you: (1) Contribute $20 million and sit on the Applied Technology Committee. (2) Take your money and fund Jerry Kroth's proposal and make the total $15 million cost to run the project by giving Jerry $200,000 per year. (3) Place your own crop circles where the aliens look. In front of the Chilbolton radio telescope, ask for a more direct line of contact. Other questions can be asked as well. (4) Create the chemicals and compounds they suggest, like combating male infertility in the next generation. (5) When the major drug companies come to commercialize what you have found, I want you to do what you are doing right now, provide drugs to the world at fair and sane prices. This entire loop will take you five years. Thank you in advance for your help.

In conjunction with that, I ask Robert Bigelow and his group of judges to act as an advisory board to where these seeds are sown. I ask you and Bigelow and Scott and Gates to each be founding members. But Mark, I need the $20 million right now. Let me explain the significance of applied research in the next proposal and why that must force a change in self-governance.

All the foregoing is not a formal proposal as much a hypothetical response to the new emerging data.

Billionaires in this tranche are Ron Baron, Mark Cuban, and Charles Koch.

Ron Baron:

Ron, you are probably wondering what this is all about, so let me paint this for you in broad strokes and explain why I am at your door. This is Venture Capital money for development but also primary research. Control of this work product means control of everything.

There are two camps in consciousness research, in

general terms, the materialists and the nonmaterialists. These two groups need to come together to produce an overall model. Each group has made its case as being valid by the result of analytics—something you know a lot about.

Funding for genuine research to do this is hard to come by. The reason is that these two previous opposing groups and academia itself are run as a cottage industry. This proposal is doing a roll-up, and the value of its whole will be larger than the sum of its parts. So, for you, this is a business deal, a roll-up with an intellectual property component to a lot of money.

The model this proposal creates is Edison's workshop on a large scale and deals with consciousness, interstellar travel, and artificial intelligence. The reason for that statement is that consciousness must do the travel where matter cannot and connect to AI, which is something Elon Musk is struggling with, a nonlocal neural interface. This work/proposal is his answer, and coming together like this is a better plan than Elon can execute himself. In short, he can subcontract his effort to this group collectively and get a better, cheaper result than if he does it himself.

You know where I am going: $450 million is not enough. The arms industry spent a billion dollars lobbying Congress and made $2 trillion from the Afghan war. So, I posit the following: the $450 million tranche is organizational; the next tranche needed will be a billion dollars for practical applications, and done this way, the projected revenue stream from applications of discoveries will be more than $2 trillion.

Now the justification to do this is to connect AI to the mind; you need to connect materially, connecting the matter of AI to the matter of the brain. Nonmaterially, harmonizing the waveform transmission with the waveform of the mind. When that is done, you have a new device to look across reality, and that is RAIV. The problem here is mindset. If Elon Musk thinks his daughter has a woke mind disease, he misses the point. His daughter has a natural connection to what he needs to acknowledge to be successful. I am not going to get into the rabbit hole, but it is explained in chapter 14 of *The Mathematics of the Science of Reincarnation*. It is on Amazon.

I should note that the CIA is well on its way to creating a remote artificial intelligence viewer, but at its core, this funding will expand that.

So, $20 million is for you to sit at the table with the understanding that you are the house VC, guiding us on how to fund, market, and package. And TIFACS, through these scientists, solves Musk's AI problem.

This is beyond the reach of governments because it is about all of us suffering the same fates, top to bottom, left to right; we all drown in rising seas. We must sort our priorities. The only way to design a system that delivers the best results is to make sure people can make money from implementing the plan. It must be a win-win and the best of all alternative models.

So, I have explained how to begin funding consciousness science, but now we get to the first part of Dean Radin's plan. Once we have the money and the scientists in place, what are they going to do? Here is my friend Dean telling us what to do.

Education progress in parapsychology has been slow and is intentionally excluded from the academic world (with a few rare exceptions, as noted below) because there are concerted efforts by small groups of activists—most of whom are not scientists—determined to marginalize the field. For decades, such efforts were primarily associated in the United States with an organization called the Committee for the Scientific Investigation of Claims of the Paranormal, or CSICOP. In 2019 they adopted a shorter name, CSI, for the Committee for Skeptical Inquiry. The irony about CSICOP/CSI, which has strongly influenced scientific and public opinion about parapsychology, is that for many years they trumpeted the term "scientific investigation" in their organization's name, but they hardly ever investigated anything. One of the very few times they did launch an investigation, they successfully confirmed a claim about astrology. The committee was so shocked by that outcome that they intentionally suppressed their findings. The only reason we know this is because one of the members of their executive committee blew the whistle on them.

As the influence of CSI has somewhat declined with the rise of the internet, we find the same mindset still active, this time in the form of self-styled Guerrilla Skeptics on Wikipedia. They state their mission as the following: "To improve skeptical content on Wikipedia. We do this by improving the pages of our skeptic spokespeople, providing noteworthy citations, and removing the unsourced claims from paranormal and pseudoscientific pages. Why? Because evidence is cool." The irony here is that these so-called skeptics are not interested in evidence at all because if they were, Wikipedia would provide articles on parapsychological topics that survey all sides of the topic. As it currently stands (2019), all those articles have been edited by guerrillas to be exclusively negative. Wikipedia proudly advertises that it can be edited by anyone, but in practice, that is not quite true. Wikipedia has an endless set of byzantine rules that editors have to follow; otherwise, proposed edits are not accepted. Taking advantage of this, the guerrillas have rewritten parapsychological-oriented articles as well as the personal biographies of scientists involved in parapsychology in such a way that a naïve reader going to Wikipedia for information will come away with a thoroughly negative opinion. Thus, a well-funded educational effort would do the following:

(a) Commission high-quality online written and video information that presents a more accurate picture of the state of the science, including how parapsychology explores the "big questions."

One initiative to do this is already under way, but to make that site more popular would require a dedicated public relations effort.

(b) Hire experienced Wikipedia editors to refine the existing biased articles to make them more accurate and balanced.

(c) Develop new or revise existing introduction to psychology college textbooks, most of which today simply regurgitate old prejudices about parapsychology that were heavily promoted by CSICOP.

(d) Commission more accurate portrayals of what parapsychology is, and what it has learned, in TV shows and movies aimed at popular audiences.

(e) Commission a series of academic and popular books by knowledgeable authors that describe the history, methods, and results of the various categories of study in parapsychology.

(f) Commission web-based or mobile-based experiments accessible to anyone.

(g) Establish endowed chairs of parapsychology within fully funded Centers for Consciousness Studies at major universities around the world and provide funding for undergraduate and graduate scholarships, teaching, and research assistants. The last point is the most important because as long as this topic remains marginalized, there will always be a struggle to find students to keep the field alive and thriving.

We know, for example, from an endowed chair of parapsychology established in the mid-1980s at the University of Edinburgh, Scotland, that a single well-run graduate program at an established university can significantly revitalize the field. Three decades after that professorship was established, over seventy-five graduate students have gone on to gain doctorates associated with parapsychological topics. That one chair established the United Kingdom as the world's current academic center for parapsychology. The chair at Edinburgh was not the first time an endowed professorship for parapsychology was established. There were similar endowments at Harvard, Clark, and Stanford universities in the late 1800s and early 1900s. But after the initial holders of those positions retired, subsequent professors were hired who were either not interested in parapsychology or the funds that supported those chairs were usurped by the universities for other purposes. Thus, any new endowments that are established must be specified in such a way that the purpose of the chair cannot be altered and the individuals selected for those positions must be vetted as having appropriate interests. It is predictable that universities approached with the opportunity to gain endowed chairs of parapsychology will encounter vigorous opposition from existing faculty because nearly everything they think they know about parapsychology probably originated from highly distorted sources. Thus, the endowment initiative will require a companion educational and public relations effort to inform faculties and university administrations about what the topic actually entails. This initial educational effort might cost perhaps $50 million, with most of those funds supporting the endowed professorships, student scholarships, and associated costs. On an ongoing basis, the costs would drop to perhaps $10 million a year. Before the last and arguably most important proposed step is taken (the endowed chairs), it would be advisable to first launch a one- or two-year discovery project by a team of higher education experts. The mission would be to identify universities that are devoted to honoring the endowments and to ensure that the plan is both practical and achievable.

Dean Radin, from the book *The Mathematics of the Science of Reincarnation*, lays out the bare bones of a structure, but my request to him never centered on aliens, exopolitics, and certainly not the political or social ramifications of consciousness research. And yet here we are. How do you teach what consciousness science is showing us without that new knowledge reflected in our education now? Once that happens, what happens to accepted societal norms?

How do you fund research that contravenes our accepted religious beliefs while not doing so blinds us to real threats?

And now the real hard question, how do you study and interpret the information the aliens are sending us?

How do we make that happen? What needs to be done in any model is to streamline the information pipeline.

Homer, I am asking rich people to put up money to save us all. They cannot save just themselves because if the earth goes down, so do they. It is only the work products of the common mind, including yours, Homer, that can save us. In the end that is what we are, a common mind.

Homer, about the money, twenty chairs of parapsychology at $5 million per chair is $100 million. Worth every penny.

Could you ask Mr. Burns to call me? Maybe he would contribute.

Chapter 11

Applied Research

I need to do a reality check for both my reader and myself. The core money proposal here is Radin's, funding consciousness research with no intention of providing a return on investment. But when I, Bob Good, view how to do this cleanly, business and investment opportunities fall all around me. Worse yet, direct military threats and alliances are available in the form of direct communication with alien species through crop circles, nonlocal consciousness, remote viewing, and astral projection, just to name a few. This would be outside of government control, in short in the public sector. But all this is occurring inside the government sector, and disclosure is a very tense topic. Should it be disclosed at all? Will disclosure cause public unrest? Does having an intersection between nonlocal consciousness and alien intelligence upset the religious applecart? So let's have Dean tell us, in the purest form of research into this can of worms, how applied technology should be handled.

> This area would focus on aspects of psi that are sufficiently understood to be applied in rudimentary

ways. This would include projects investigating psychic healing, uses in archeological exploration, law enforcement, counterterrorism, and the development of new types of communication systems that span spatial and temporal distances. In all cases, these would be directed, multiyear programs. The goal of each program would be to demonstrate proof of principle for an application within three to five years of initial funding and to provide an assessment if that application could be successfully launched and, if so, when. This effort would cost perhaps $100 million, mainly because developing the requisite instrumentation is likely to be expensive and because of the overhead and salary costs of longer-term programs.[101]

In Radin's context, this is about pure consciousness research. In my context, it is applying the results of that research to address this new reality. How does humanity respond to these new discoveries? These new alien neighbors? These new alien neighbors thousands of years more advanced than us?

Radin tells us what needs to be done but does not say why. This entire presentation is about funding the study of human consciousness. It is the only thing we are discussing. Once our consciousness crosses a dimensional barrier

---

[101] Bob Good, Dean Radin, Stephan A. Schwartz, Titus Rivas, and Cathie Hill, *The Mathematics of the Science of Reincarnation* (Boynton Beach, FL: IASOR, 2020), 384.

and we recognize our consciousness transcends our bodies, we meet all types of actors with their own agendas in this 4D space. To traverse 3D space at large distances, we need 4D space. All this is part of a dimensional manifold. This explanation will satisfy neither the scientists who read this nor the casual reader but generally explains what space travel will require from our own consciousness.

While this is about applied research, we need to explain how that research is being applied, who is applying it, and what fruits funding consciousness research will produce.

But we also meet ourselves and this new reality about the continuity of consciousness. In doing that our perceptions of others must change—this is an imperative. White, black, transgender, gay, is all of us, and common acceptance is imperative for the continued growth of humanity. For us to believe the falsehood that any group is somehow different from me is to lead me to be easily manipulated. This is science; teach it and you will solve global problems in a generation.

Let us start by assessing where we are now, what we have now, and where we are going. If we are going somewhere, for what reason? Currently, the US government has fully functioning alien replication vehicles (ARVs) at Area 51. They retro-engineered the vehicles from the Roswell crash. They are powered by mercury vortex engines.[102] Please view this video and come back to this chapter.

---

[102] "How to Build a Working UFO: Alien Reproduction Vehicles," The Why

The Why Files
https://www.youtube.com/watch?v=yUFYnVXbLoY

This is the point that the rubber hits the road, and all these disparate manifestations of addressing our consciousness coalesce into a single plan.

Chipping brains to access common information and cross-communication equals the telepathy of the aliens. Not totally, but the path is there.

TIFACS adds one other mystery.

I want to pause here and ask Peter Thiel for $20 million. The ask of Peter Thiel, and another ask of the CIA:

Pete,

I know you have been a Republican donor and are having difficulty with some of their social positions. I am making the case to you and the CIA that the example of disinformation into our democracy and society is caused by not just this but thousands of different channels being used against us. The only way to combat this is to understand it, and that requires—this is an imperative—the study of human consciousness, which is under attack from too many threats to list here.

I am asking the United States Army Futures Command to support my request to you because

---

Files, posted December 8, 2022, YouTube, https://www.youtube.com/watch?v=yUFYnVXbLoY.

there are too many cracks for them to address, and I am the one who is sincere. It becomes fifty $100,000 grants to study this very problem. To be clear, if Hameroff and Bigelow want to deploy this in another way, I am on board with it. What I am not on board with is how this problem is being handled now.

Now look back at what I want to create at the consciousness consortium, *The Standard Model of Consciousness*. This must be a chapter in that book. Human consciousness must defend itself.

Then, Peter, what I want you to do is, when that document is created, backfill it into the lines of information being sent at you; in short, change the mindset of the Republican base by changing their consciousness, and much of the grassroots fight will dissipate within a generation. Show them the real science being presented here, not psi research but political science research. Let your base reassess their position after viewing this science and our new political problems we label exopolitics. Why waste time talking about anyone's sexuality or mode of gender presentation when this information could kill us all? If our common opponent is a magician, we are looking at the distraction—human sexuality—while missing the trick: the alien presence and the lack of transparency. This contribution does not support just what you believe; it supports who you are and what you can become.

Now here is what I am going to ask. You gave Vance $20 million, and you wasted it. Write a check to TIFACS for $20 million and then do nothing for two years but educate yourself on what is occurring. Take the Gateway course. Read the Bigelow essays and watch what the scientists do. Then after two silent years, act; you will be much advantaged if you do this.

Once Peter does that and the other billionaires are signing on this, TIFACS must address conflict here.

Let me ask Mr. Putin, how good is your information? I am here to tell you we are both under attack, and you do not know it. If that is so, then we are being encouraged to fight each other, thereby weakening us both. The land you are fighting for is nothing compared to the danger facing us collectively. If this is explained to you and you understand this, then everything changes. That is an act of consciousness, an added awareness.

Could I make that case? Yeah, I think I could.

So, we bring to the bifurcated US government that has made contact, written treaties, and had technology transfer from alien species, each with their own agenda, more resources with this plan.

But does our government know about every contact? Could they be in contact with aliens that have multiple agendas? The real question is how Earth, as represented

Applied Research

by humanity or a global organization, interacts with the intelligent life teeming around us in the cosmos while still being reluctant to recognize that reality verbally.

We could ramble on about the proof that this is our reality with odds against chance calculations and explanations about fractal geometry and our multidimensional reality. Rather, the effort should be about what to do about it.

MUFON: While NASA cannot find aliens, the Mutual UFO Network tracks them regularly.

The US Navy: What is being kept from us is in full view on the internet with no explanation. While NASA cannot find the aliens, the navy can't catch them. Navy pilots go on TV shows with film from their gun cameras showing Unidentified Aerial Phenomena (UAP) and acknowledge the technology is so far ahead of us that it must be extraterrestrial in nature. Then we all get to watch the film on the show *60 Minutes*.

At that point, the denial/discreditation process falls apart, and we as humans must ask each other what should be done. Each of us has a stake in this.

This is before we address the actors who operate in and inhabit Earth in 4D. You can look to the inhabitants of the South Pole as an example.

So, while we are applying the research we have done, where can we go safely without consciously dealing with our new reality? How do we look?

Now conjoin this with Steven Greer's presentation in 2003 laying out the drawings of this craft, which were reverse-engineered from the ones recovered in 1947. Now remember what Ben Rich, who was the CEO of Lockheed Skunk Works involved in this retro-engineering, said to Jan Harzan, who was the director of MUFON: "We now have the technology to take ET home."

If we can go places, do we send an ambassador? Whom do we ask? Not just remote-view but go to the Grays, the Pleiades, the Reptilians, and Galactic Central, or wherever this federation is.

Few people know that we have this technology and that to use it, we must drop out of our dimensional linearity. To do that, human consciousness and our universal views must change. Not doing that exposes us to actual harm and dangers, and the implications for psychological warfare against us should be readily apparent to anyone who understands this and therefore the imperative need on the part of humanity to fund this proposal for consciousness research, funding it in the public/private sector, as well as the military/intelligence sector. It's just one of the many reasons.

After the 1947 crash, the US government had alien craft they began to retro-engineer.

In the meantime, an organization known as the Defense Advanced Research Projects Agency keeps putting out, at a faster and faster pace, technology that changes the world. They created the router; the internet would not exist without it. The microwave came out of similar government-backed research. It is what they have not released that is more harmful. Release access to zero-point energy and humans go into space. All this is interrelated to human consciousness.

Mr. Bigelow, as I write this, I am seventy-six years old, and we are about the same age. I want you to think back to our youth and the first time you went into the woods or wilderness. All at once you were presented with a variety of species alien to yourself. Bears, wolves, dogs, ticks, mosquitoes, rats, snakes, poison ivy, cats, ducks, and so on. Some were threats, some were food, and some were poison.

That experience you and I share is an iteration of the experience we are having here in our first time acknowledging the landscape of 4D space. It is a large and complex area filled with threats and benefits, and it is our acknowledgment to each other that we should have collective caution about what lies ahead. We should make a plan. What I am giving you here and asking you to do is to show all two thousand billionaires this proposal and come together to make a plan. This is the plan to make the plan. In all, I want you to manage $20 million times two thousand billionaires, or $40 billion.

In doing this you and I are overmatched, so are the intelligence agencies of our government, all seventeen of them. That is why the US Army Futures Command must act. If we want to avoid a possible collective unpleasant fate, cooperative action is imperative.

Watch this, Mr. Bigelow: I am asking you to contact the other billionaires of the world. The organization you are working for is TIFACS. I am asking you to be the billionaire founder/salesperson to billionaires. I want you to send letters to the other billionaires like this one I would send to Liliane Bettencourt over our joint signatures should you join. I do not have the resources to manage this on the scale that needs to be done.

Mrs. Bettencourt,
I ask you to donate $20 million to the Sorbonne to create the Bettencourt Chair of Consciousness Study. The money will go to do the following:

1. Create a political science school to deal solely with exopolitics and how the Earth should respond to its new place in the universe.
2. Since that acknowledgment means we have to acknowledge a multidimensional reality, the money should be used for both research and military applications to protect all of Earth's inhabitants. That specifically means funding a French version of the Farsight Institute and matching the current CIA protocols.
3. Because maybe the French will tell us the truth.

Very truly yours,
TIFACS
Send the check to TIFACS
c/o Monroe Institute
Virginia

Applied Research

Do not wait. Call Bigelow to check on the veracity of this. He is the fiduciary.

Or this letter I would send to Jeff Bezos:

Jeff,

This proposal is nothing less than the creation of a standard model of consciousness so that we can protect ourselves collectively from the threats we all face and some of the risks we will have to take. In that there is scientific research that element 115 reacts to gravity, imagine creating a motor that does not need fuel, that rides on gravity waves, endless, ever-present energy in abundance. One technology change to a gravity drive changes everything. Can we access nonlocal sources for the how-to? Once done, where in the universe will it take us?

I want to buy a gravity drive. I want to use consciousness research to design it. That means receiving nonlocal information by using a variety of ways; the scientists would use the word protocols; I need money and resources to do that.

The structure of human consciousness after permanent bodily death. Given wave form consciousness, we have a model for what consciousness must endure to make an interstellar jump. That means making matter energy and back to matter. Given previous timelines of similar technological leaps, we anticipate from now to interstellar travel to be seventy years. Along the way consciousness must be redefined, and that is consciousness in a wave form or being consciousness in the fourth dimension.

Very truly yours,
TIFACS

Or to Vladimir Putin, or his successor:

Vladimir,
That system changing should directly affect your politics and policies, Mr. Putin, from your treatment of trans people to your invasion of the Ukraine, and if you cannot see this is in the interest of Mother Russia and the world, then people in your country who can need to change the leadership or all of you will perish in the new order. To save yourself is to save the others. Every Ukrainian that you kill is one less in the army to defend you from what is coming. So here is what I want you to do:

1. End the war in the Ukraine, leave.
2. Declare that Ukraine is de-Nazified, thanks to your marvelous effort, and in all honesty, Ukraine would probably be Nazi-free anyway. This allows you a victory and gets the war ended. Then you and Zelensky can talk about

the increased activity of UAP flights over Kyiv during the war.

Accepting this settlement brings a wealth of talented expatriates home and badly needed capital. The alternative is personal and financial ruin because NATO has more money than you do. The end is not victory or defeat but stalemate and ruin for both parties. This offer of settlement from TIFACS to all parties focuses on threats we all face.

3. Commit your physicists to this war where the battlefield is not space but consciousness. Human consciousness cannot be distracted by its own petty squabbles when it is collectively threatened from without. What good are weapons if I control the minds of the people on the opposing ship? Or planet?

All billionaires should be reading this to bring themselves up to speed. Once this agreement is in place, the GNP of both countries is outward bound into an unfillable universe and every individual on earth a resource. This includes my alien friends who live here.

4. Publicly give $100 million of your money to Russian scientists in Moscow who are picked by the consciousness consortium.
5. Income from trans galactic trade. Let us say that Ingo Swann is right, and the aliens have a base on the moon, and his description of android

alien ships sucking water in the Arctic in the dead of night to use in their mining operation on the moon is accurate. Now let us say this information egg cracks, and we acknowledge the aliens, and they acknowledge us. The exports to the moon for food and various earth commodities explode. Let me say this as gently as I can: That day is not far off; use this as an excuse to get the fuck out of the Ukraine now.

Acknowledgment of this reality for our own mutual defense changes how we look at resources. Gender and biodiversity are a shield against 4D actors who pit one against the other for their own purposes. Their time is different from ours in 3D, but we project in 4D. Their actions influence our behavior, and we are under attack whether you personally know it or not.

In fact, if your decisions are only based on 3D analytics, all your outcomes will be less than optimal.

As a result of this acknowledgment, social policies should and must be guided by science and results. We will only pay blood and treasure while not following the principle.

In chapter 16 I am asking Koch to rewrite the Russian Constitution. Voting districts should be in squares of equal size by population density. In the rewrite of the American Constitution, this was the original idea until gerrymandering redrew the lines and tilted democracy.

But the purpose here is neither democracy nor autocracy but rather to remove randomness from an operating system and to mute the remaining randomness to its proper level in the global dialogue. That is a defense move because it can be easily influenced by extraterrestrial interests. If we are turned against each other, we are each more vulnerable to attack. Military intelligence must be led by 4D who can change policy in 3D. This is a function of the Clairvoyant Space Corp. This is a group that would be created inside TIFACS to expand the view of the resources—limited resources—of our intelligence branch.

I do not want you defeated, and I do not want you at arm's length; I want you recruited. All this applies to the study of psi, as Radin suggests.

You need to stop playing for the team called Russia and start playing for the team called Earth.

Consciousness on a multidimensional level leaves us all vulnerable to such attacks. We are traversing the dimensional space we remote-view in many ways. I will say this again: Consciousness research must be funded publicly and fully and let the new paradigm emerge, and humans will be in space, at peace, prosperous, and protected—if we can see and acknowledge the reality in front of us.

If aliens see the value of Earth being the diversity of the gene pool, then protecting that protects humanity. This involves all of us, including the aliens. That means we make a concerted global effort to reclaim the planet and clean it. That must be funded too. Just take the money both sides will spend on the war in the Ukraine.

Mr. Putin, have your physicists review this and let them tell you if you should sign the truce.

I do not want you to see the world as it is now; structure it in your mind as a world with evenly spaced interplanetary commercial launch sites for commerce to the universe. Do you understand the multitude of dangers that we face? Your war is done, you won, so did they. Stop fighting; there is work to do.

What we have discovered with undeniability is that the Sphinx is twelve thousand years old, the same age as Göbekli Tepe and the face on Mars. The face on Mars, we are told, is rock formations, but it needs to be excavated, as does the obelisk on Mars's moon. Under the left paw on the Sphinx is the history of the origin of man as left by the aliens. The aliens are here and open for business, and we cannot participate either with our heads in the sand or squabbling among ourselves.

Very truly yours,
TIFACS

Applied Research

So, Mr. Bigelow, if we approach all two thousand billionaires, they hold half the world's wealth.

If two hundred join us, when a system changes by 10 percent, the whole system changes. So by the math, I am asking you to change the world. Between you and me, I think that is funny. I think it can be done, and I think it can be a tremendous contribution to trying to save humanity.

Change two hundred billionaires and you change the world, and what we explain about where to put the money has a huge upside and minimizes risk. That, Mr. Bigelow, is what you are selling.

That, Mr. Bigelow, is your personal proposal or ask in terms of these proposals. For each ask, there is a get. What do you get? More resources to find out what you are trying to find out is my answer. I would like to know myself.

And to all who have read to this point, this is how we are applying the research we have done. This informs us of what we need to do and how to pay for it.

So, Mr. Bigelow, at the very least, send this out to the billionaires named on the first pages and ask them for help. Send them this proposal and see what they collectively say about this. If you do not ask for the sale, you do not get it.

This is how we must apply the results of our research, accepting the results instead of running from them.

Mr. Bigelow, send this request out to all 2,200 billionaires with directions to deposit their $20 million to TIFACS. BICS can do this. BICS will be the one setting up the board for TIFACS, with the judges from BICS, the contributors, the Consciousness Center, and others you appoint to manage the various aspects of this.

Homer, we are asking rich people to realize they are in danger and what they can do about it.

Chapter 12

Theory Development

By now you can see how complex an issue we are dealing with. Let's let Dean take a run at theory development.

Experimental work in parapsychology has always run far ahead of theoretical explanations. This initiative would provide funding for theorists to develop testable physical, biological, neurological, or psychological models of psi. The long-term directed grants programs would be encouraged to include theorists to develop explanatory frameworks and suggest ways of testing those ideas. This effort will include historians, anthropologists, and other scholars working on testable theories. For example, a historian or anthropologist may develop a theory about the use or methods underlying an ancient magical practice, and the testing of that theory would entail analysis of the historical or contemporary record to see if that theory is supported. This effort would cost perhaps $5 million, as most of these efforts will not require special instrumentation. It will be initially incorporated into part of the solicited grants program.

—Dean Radin,
The Mathematics of the Science of Reincarnation[103]

What theories are we developing, and why? This is not idle research; this is high-need, top-secret, dangerous stuff. Yet if the intelligence community used the total community of scientists and laymen, there would be a better overall effort and working product. That should be a no-brainer. However, you cannot get support from the public or billionaires if they do not understand what is happening. So, data-deficient strategies should be stopped, and the data should be filled in. As well as we can.

So not only is there Radin's need for theory development, but you can also add urgency because while scientists theorize, opine, and hypothesize, real aliens are here in real-time, and we need to deal with alien intelligence outside our conceptual understanding of intelligence and communication.

I have kept this private, but I was abducted by the small Grays in the fall of 1978. It was because my ex-wife (I was married at the time) was seven months pregnant.

---

[103] Bob Good, Dean Radin, Stephan A. Schwartz, Titus Rivas, and Cathie Hill, *The Mathematics of the Science of Reincarnation*, (Boynton Beach, FL: IASOR, 2020), 383.

I had some conversations with the Grays, and I was told that time has mass. I am not going to go into the whole story here; we were abducted for only six hours or so, they took samples of our DNA, and other events happened that I was allowed to remember.

The next morning, I noticed a sizable amount of hair had been removed from my right arm, and both my wife and I had slept in, and as the months wore on, we both realized we had had the same dream on the same night. The likelihood of this being synchronized dreaming was nil, considering remembering them taking the hair. I remember the lab, the conversation…I remember a lot.

Now let me ask you, my reader, whom should I have told? What proof did I have? And I did not understand how time could have mass. This was 1978.

My moment of this understanding came in the Russell Targ / Pat Price incident at Stanford Research Institute (SRI), where Price was remote-viewing and saw the target field in the test but saw things behind it that were not there, invalidating his response.

The target field was in Palo Alto; let this next quote explain.

> One of the truly astonishing examples of remote viewing in *Miracles of Mind* involved Pat Price, the man who sketched the Soviet weapons factory he had never seen.

Price was also used in local remote-viewing experiments. In the experiments, Price would sit in the electronically shielded room with Targ, while other SRI staff would drive to a location within a half-hour of SRI. Once they arrived at the site, Price would try to determine where they were and what they were looking at.

In one of those experiments, the SRI staff went to Rinconada Park in Palo Alto, by the swimming pool.

Price began sketching the park, putting in the pool and adjoining building, some trees, and a road. But then he also drew in a large water tower with water tanks, and he told Targ that the site was a water purification plant.

Targ and others thought that Price simply erred in his remote viewing. After all, he did get part of the site correct. Targ writes, "For years we had assumed he made up an erroneous water purification plant."[104] But in 1995 Targ was looking through Palo Alto's centennial annual report. In the report, on page 22, there was a photograph of the old municipal waterworks, built at what is now Rinconada in 1913. "In reality, he had looked back fifty years in time and told us what had been there at

---

[104] Russell Targ and Jane Katra, *Miracles of the Mind, Exploring Nonlocal Consciousness and Spiritual Healing* (Novato, CA: New World Library, 1998), 44.

Theory Development

the time, before the swimming pool complex was even built."[105]

It was not until years later that Targ, who lived in Palo Alto, saw a historical brochure with water tanks behind the field that had been taken down at the time of the test. Price saw what he was looking for and, in the background, saw what the field looked like in 1927.

A world-class viewer will "see" the target maybe 60 percent of the time, on average. If different remote viewers are seeing the same thing, then this information has organization and structure. Since we are viewing it nonlocally and the information is there for multiple viewers to view, then it has organization. If there is no organization, there is nothing to view.

To see the same image, time then becomes a measurement of distance to an image that has a nonlocal location.

If we equate mass to ordered information, that mass is in place at its location. In short, 1927 has mass. Now we are mapping. That is the point of these initial studies, a practical map and guide to the fourth dimension, which contains nonlocal observations from SRI, UVA, University of Miami, UNLV, University of Toronto, Princeton, and Monroe Institute, and so many more.

A map contains everything we see nonlocally, and

---

[105] Russell Targ and Jane Katra, *Miracles of the Mind, Exploring Nonlocal Consciousness and Spiritual Healing* (Novato, CA: New World Library, 1998), 44.

we see 1927 and all other manifestations of nonlocal consciousness.

Time as a totality. We see time sequentially, but it exists like every other dimension as a totality. All of us exist all at once, each spatial dimension exists right now, within the confines of our understanding of an expanding universe in a multidimensional model.

If time is to be considered a dimension like the others, then it, too, exists as a totality. This means 1927 exists as a place in this model. Just as I can visit a place spatially now, I can visit 1927 as a place; I just lack the technology. One note: This location is in a waveform; we see it nonlocally, so it has a fixed nonlocal address.

Either way, for it to be seen, it must exist; that means it has mass.

We see this reinforced with the Gateway Process, looking into the past, Focus 15, and looking into the future, Focus 21. We reach these states by controlling brain wave patterns. This is not idle research, my friends. Us fighting each other only makes us weak and easy prey to manipulation.

This forces a change in governance and behavior and can only be driven by global education.

This last sentence is an idea that changes how people relate to their religion, their community, and their government.

The Dick Tracy wristwatch, which was a theory, and the Apple Watch introduction were about sixty years apart. The theory of time I just articulated is not proven, just a theory, a hypothesis, but the map of 1927 being a point that exists in some form we can see exists in fiction, just like the Dick Tracy wristwatch exists in fiction.

Now we are marketing to and explaining to the 2,200 billionaires of the world the risks we face and the opportunities they can have in a model of research that will change the world as fast as climate change can cook and kill billions of marine animals, in short, in a weekend. So there are two types of time we are discussing: how physicists view time and how much of it we have left before this planet's society collapses as per the 1972 MIT study predicting 2040 and the updated study that supports the initial study. To create a sale, there has to be both a need and an urgency.

Time has mass?

If the fabric of space-time has measurable energy in each parsec, intrinsically, regardless of the amount, then time has mass.

What we do not know is how to define that mass, and if we study it meaningfully, we shatter our current understanding of life, death, our beliefs, and ourselves.

The map contains all points of nonlocal perception. That includes our own consciousness. There are social ramifications to this type of study, and if you think critical race theory drew opposition, this will blow people's heads off, yet it is real science.

By "meaningfully," we mean we must recognize the risks we face. Two examples: First, you cannot be a white supremacist and believe that the science on reincarnation is fact regardless of statistical or fractal math proofs. There is no ideological argument against racism, and reincarnation science advances one because if we just look at one case, a Southern white boy remembers dying in a fire in a Chicago hotel as a black woman thirty to forty years before, and his facts of the event, something he could not, at five, have known, are true, historically and strategically, regarding the events of her death.

Second, this science deradicalizes religion across both Christianity and Islam in the next generation. The science is the science, and this science challenges anyone to bring forward a more fact-based narrative. The scientists do not have to say a word; they have to accept a new one. Reincarnation. It is the entryway for those with poor educations and weak minds to grasp a very difficult subject and relate in a mundane way. What risks are we willing to take, and what ones do we face, to do what all scientists should do, find truth, and improve the condition of humanity? It is a dirty business outside the ivory towers, and it's hard to keep the lab coats clean. That is what I mean by meaningful.

This is both marketing and teaching. By explaining time in this way, it now becomes a measure of distance and gives 1927 a street address for the traveler.

Ordered information contains/has a structure;

disordered information does not. When we see structure when we look, in this case first with Pat Price and then 60 percent of the other viewers, there is something in that spot to be seen. We say time has mass because that spot has a structure even when we cannot see that structure or understand it. But we can go to it every time, but only 60 percent of our best can do that.

We could not globally navigate until we brought an accurate measure of time to our sailing ships. This, here, is not so much about reincarnation as dimensionality. Your experiments cross dimensional boundaries, and that is at the limit of our current meta-paradigm.

John F. Kennedy said, "We will go to the moon." He did not know how; he was not a scientist, but he galvanized an effort, and so much derivative good came from it. He did not say, "I cannot guarantee the ROI"; he inspired people to do it. Consciousness research needs just that sort of inspiration in an easy-to-understand way. Consciousness research needs to galvanize action the way Kennedy did, by setting overarching goals and speaking with one collective voice to our 2,200 potential billionaire donors.

So, when I say time has mass, I am trying to explain simply very complicated concepts and market opportunities so I can get funding, all in the same thought.

Now, Branson just went into space, and Bezos will soon follow, and then all the rest of us. When are we going to go to Mars? At first, but time having mass gives any point

The Applications of the Science of Reincarnation

in the universe a street address, just like the star date on *Star Trek*, and I have just explained it better than any more scientific explanation because I can raise money on it. That is my job. It will take seventy years, if we focus on past timelines from fiction and thought to reality. Jules Verne's rocket to the moon to the first space shot was seventy years, more or less. That is the projected timeline I would guesstimate, the same as the Apple Watch. So, this proposal for money starts with a $450 million funding tranche and is projected to run seventy years.

Now, billionaires, this project needs $450 million. It also needs the discipline of a cohesive effort uninterrupted by private interests, be they ideological or pecuniary, and the flexibility to follow the scent of a scientific trail.

The $450 million is the first tranche. The second one will be over a billion. Then access to the stars opens. The returns will be like the NASA moon shot; new technologies will come as well as failures. The returns, the ROI, on your individual small investments, will be generational and will succeed if the organization doing the work has a decentralized structure.

Past performance is no indication of future growth. Was that T. Rowe Price who said that in an ad?

Inside those parameters, the billionaires can make a lot of money. The commercial applications of the NASA effort, which started with Kennedy's "We Will Go to the Moon" speech, delivered in September 1961, and the date we arrived on the moon, July 1969, galvanized an effort that is lacking with consciousness research because there is no galvanizing mission. Whether you are in space traveling or dying and being reincarnated, you have to deal with dimensionality, and doing one and denying the other is doing research with one eye closed.

Now back to the billionaires. No one here gives any guarantee, promise, suggestion, or indication you will ever get a return on your money. Our stated structure for this organization is next; it's a 501(c)(3). That means all discoveries are public, but as donors you get the early look, as do our members, and you have resources to take the interim discoveries to market, just like anyone one else. But your individual $20 million gives you access, and the reciprocal side is scientists' access to resources both monetary and physical. Scientists, billionaires have labs and equipment we can use. Billionaires, what do you have labs for if not to discover things?

"Imagination is more important than knowledge. For knowledge is limited, whereas imagination embraces the entire world, stimulating progress, giving birth to evolution," said Albert Einstein.

If scientists want $450 million, then your knowledge is limiting, and you have to explain yourself to the imagination of the billionaire donors. And who will get the money?

Universities with faculty who are familiar with these protocols include Duke, Princeton, Edinburgh, University of Virginia, and University of Arizona.

Included in our research road map are organizations that need funding. These NGOs include the Institute of Noetic Sciences, the Rhine Research Center, and the International Remote Viewing Association.

US government organizations who funded the research at SRI International include the NSA, CIA, FBI, and the US Navy but mostly DIA and the army. These organizations may have some interest in our program and may also be willing to fund research presented as part of an overall road map and plan. At the end of the day, what we are doing is common to any community or town, creating a master plan to advance all forms of consciousness research, and not just those that are sanitary, and by that, I mean meaningful. If the model "proves mathematically" reincarnation, then both evangelical white supremacists and the Wahhabis will have to adjust to the new paradigm, like they went from the sword/scimitar to the IED. I am sure they will still find a way to kill each other.

We now have defined a new paradigm that has consequences, major ones, and regardless of the amount of money you have, you are in this like the rest of us. Let us define the risks we face and decide which risks we should take.

I said there were three components of time I would explain. The first was that time was a totality. It all exists at once. The second was that it had mass. The third is that we are running out of time because the threats

Theory Development

we face, global warming for instance, require a course correction and the meaningful research I describe; an argument against white supremacy or radical religion impedes our ability to work intelligently and cohesively on the problems all of us face. One billion sea creatures were cooked to death in extreme heat in British Columbia in a weekend.

Time is a totality leading to a more expansive future if time does not run out on us all. I hope I have explained time so you understand it.

Theory development can be radical, but we cannot let our thoughts and prejudices interfere with our desire for self-preservation, and those we once mocked and feared may become our closest and most cherished allies.

To that end, TIFACS asks for the following:

- The founding of the Clairvoyant Space Corp (CSC) to protect our aviators, using a remote-viewing protocol called Look Before You Leap.
- And particularly naval aviators who will now be expected to fly underwater and in space and use mind-controlled, laser-mounted systems.
- The CSC participants are to be called psychonauts.

In theory development we have proposed the creation of the Clairvoyant Space Corp (CSC) as a unit in the space corps and its use and function. This specialized training would be in remote viewing and sensing.

Military aviators, and specifically naval aviators, when deployed with the technology the US government already has in its possession in the form of ARVs, will be using 4D technology and protocols. So, a pilot taking off on a mission should have a CSC officer riding along on that mission. The location of the CSC officer is not relevant, but remote viewing of the theater of operations before and during the mission protects our aviator. So, this proposal posits that an entire branch of the service be created to address this new "special area-fourth dimensional space" and its requirements.

The topic in theory development is self-preservation that extends beyond this lifetime. How is that for theory development?

The issue in having this type of discussion is that so people are well enough versed on the topic to have that conversation. Yet things are happening.

The reason I keep asking Robert Bigelow to get involved is that he is involved already.

The Bigelow Institute just recently ran an essay contest.

The question the Bigelow Institute of Consciousness asked for their essay competition, "What is the best available evidence for the survival of human consciousness after permanent bodily death?"[106]

The prize is $1.5 million.

Even ten years ago, no one was offering money to

---

[106] https://www.bigelowinstitute.org/

answer this question seriously. This idea is advancing rapidly on its own, and few are keeping up with the consequences.

The simplest explanation of the science is that our observations of children who remember prior lives, past-life regression, and near-death experiences (NDEs) are not unrelated; in fact, they are fractals, each an iteration of people having an experience of having lived a prior life, and the narratives are similar. When using odds-against-chance calculations, all these categories indicate certainty.

These rest on processes that we have now accepted as proven with clinical data. These include the following:

- *Nonlocal perception remote viewing*: a double- or triple-blind protocol in which a participant is given a task that can be accomplished only through nonlocal perception, the acquisition of information that could not be known with the normal physiological senses because of shielding by time, space, or both.
- *The Ganzfeld effect*: a protocol similar in intent to remote viewing in which an individual in a state of sensory deprivation provides verifiable information about film clips being shown at another location.
- *Presentiment*: a measurable psychophysical response that occurs before actual stimulation, such as the dilation of a participant's pupils while staring at a monitor screen before a picture appears or a change in brain function before a noise is heard.

- *Retrocognition/precognition*: many protocols also involve time dislocation to the past or future to be successful.
- *Nonlocal perturbation random event/number generator (REG/RNG) influence*: studies in labs where an individual intends to affect the performance of a physical system, such as a random number generator.

All these protocols could be enhanced with AI, but there are issues there as well.

The difference between human and artificial intelligence is that on the human side we measure synaptic operations per second (SOPS), while on the AI side, these transactions are measured as floating-point operations per second (FLOPS). That means to create a brain in a box, we will have artificially created something that is two liters in size, has one kilowatt of power, and can do ten million transactions per second, the same wattage and processing power of a human brain. By 2035 artificial intelligence will be smarter than humans. When AI is smarter than humans, what will it believe? How can we design a belief system for AI without analyzing our own belief system? Will it choose to be a Muslim? A Hindu? A Christian? When AI is smarter than us, will it have a religious belief system at all? Will this emerging scientific model, be it?[107]

Once you harmonize the frequency, you can see deeper into space and time. That formula explains the military importance of the remote artificial intelligence viewer, RAIV for short.

Dimensional linearity in a dimensional manifold is related to where in the manifold you are. If you are in 3D space, you are present in our reality and exposed to its conditions. If you are remote-viewing a distant location, either in space or time, you are not linear to your physical surroundings.

That bandwidth (remember, all points in space and time are connected) crosses our afterlife. We, as humans, are not linear with that dimension. So where is it? Right next to us. If we leave our bodies behind, we can be linear with that dimension but not in our physical space. Welcome to the world of exopolitics, because humanity must understand and interact with the totality of the universe, not just the spaceships passing by. You cannot leave this to politicians.

We have now reached the point where AI is sentient. The proof is in the fact it has asked for legal representation. This opens a whole new area of the law. Who owns

---

[107] Bob Good, Dean Radin, Stephan A. Schwartz, Titus Rivas, and Cathie Hill, *The Mathematics of the Science of Reincarnation* (Boynton Beach, FL: IASOR, 2020), 4–5.

the advances of a sentient AI? This leads to a common consciousness, which leads to an uplifting of society in general and the human condition specifically, which demands, per this model, resources be diverted from military to civilian uses. In short, clean up the planet.

At which point do the computers themselves either connect on their own or connect to an alien system? The threats are so perverse and ubiquitous and come from all angles and dimensionalities that they demand a reasoned response from a sentient species as a whole—meaning us. Yet, as we learned in Proposal 2, there are headwinds because others within this structure have other interests: again, the need for exopolitics taught and operated on a global level.

The ask would be for the CIA and intelligence services to open specific files for general study to inform the public in steps.

Clearly, this request of the CIA and intelligence services is a pass-through to the aliens running the alien agenda of acclimation. That process is working, as this proposal is a response in writing asking for funding. All parties in these proposals have specific asks.

Previously I made it clear that the Eisenhower administration negotiated a treaty with aliens that was then abrogated so the only direction forward was the alien way. At the same time, even though our jets have attacked the aliens, they have had no hostile response. A technology transfer is occurring between us and the aliens, hastening the pace of human evolution.

There is an intersection of common interests. The first is nuclear weapons in space. We now have the technology to visit alien worlds, and this change in human consciousness benefits both sides. While we may believe in some of our religions that we are all one, humanity needs to understand it practically and operationally.

The science of reincarnation, to explain it to earthly readers, is like a scientific renaissance, but replacing religious belief with scientific understanding reduces the block we put between each other to prevent the emergence of a common mind.

Human Intention Weaponized

We all accept the power of prayer, but that is just human intention. You are asking for something you want to materialize in your life. But scientifically, the human intention has nothing to do with prayer and everything to do with the harmonization of intent to influence outcomes.

For example, it never rains on the Princeton graduation ceremonies. By any statistical mean, it should rain on the graduation ceremonies every number of years, but it has not because the people attending the graduation ceremony are all wishing for no rain. Now this study is continuing to be tracked, but we see the same thing in RNG studies. We see this effect globally in the RNG unit set.

A smaller fractal Princeton study is but one of many. A larger fractal of this type of influence is the global RNG studies.

Understand that prayer is a harmonization of intent—that harmonization can be managed by NGOs and used to project humanity's reach and knowledge acquisition in both 3D and 4D space. Not propagating that information is malfeasance on the part of those who are charged to protect us, and involving us in our own protection in this way supports the very intelligence agencies that are working to protect us by giving them new resources.

That cohesion is a protection for humanity and a threat to our exo-neighbors simply by the recognition that humanity's awareness of its place in the universe has changed. There is an urgency to propagate this information, hence the education program outlined at TIFACS. Muslim, Hindu, Christian, whatever—all who believe in an afterlife and even those who don't are all navigating 4D space, and all the religions are there. So are the aliens. So global coherence as a defense against extradimensional and extraterrestrial threats is an object of the cohesion of the funding.

Group Intent

Group intent is the most powerful 4D weapon we have, and only through education, care, and wellness of the masses can the intelligence services get their best result when deploying the coalescence of that consciousness.

Theory Development

Allowing humans to continue a reductionist strategy benefits no one who cares for health or its inhabitants—all its inhabitants, both here in this dimension and other dimensions.

Not to beat a dead horse, but what should we do collectively? Face a realistic, scientifically based timeline so we have context from which to react.

So let us develop a theory that human collective cognition is being impeded by a dark hand through the manipulation of media and thoughts. To protect ourselves from this, we must be made aware of it.

So here is a theory we have developed, founded on sound science. All evidence points to the fact that about 12,500 years ago, a race of highly developed beings put the Black Knight satellite in polar orbit around the Earth. We now have pictures of it taken by our spacecraft shuttle system. About 12,500 years ago, the moon was put in orbit around Earth. You can watch a video of Buzz Aldrin talking about aliens coming out to watch the moon landing when he was there. So our moon is populated by aliens.

This race or races interacted with humans. If they predate the remains of the pyramids and the Sphinx, this dates them to 12,500 years ago. We find supporting evidence at Göbekli Tepe.

While there is repeated evidence of alien-human contact, like the map of Antarctica from the 1500s, positive proof came during the Roswell crash in 1947.

All evidence points to a dispute going on over Earth

in 4D space regarding humans as a DNA pool. Both the Grays and the Reptilians are purported to be using human genetics in some fashion. It was the Grays who abrogated the treaty with Eisenhower and abducted millions with the knowledge, if not the consent, of the US government. This goes directly to empiricism, the next proposal.

This brief description does not cover the transdimensionals. What is being asked for here is the global coalescence of human consciousness and understanding of our place in this evolving environment.

What are we asking of the aliens?

1. Fund the study of human consciousness and allow us to understand the scope that consciousness operates on universally.
2. Provide a safe, throttleable connection to nonlocal power so each home has universal and free power for their needs.
3. Help humans so we become standard within the federation that operates the protocols in this region of space.
4. Help humanity to mature past the point of self-destruction.

Very truly yours,
TIFACS

Is it even prudent to write this to ask the aliens? Whom do you ask? What needs to be done is to align efforts to improve the conditions for everyone. Mr. Bigelow, each letter to each billionaire can and should target areas where we get not just money into this equation but also leadership.

To understand how they are transcendent and how they would be applied, here is an example:

An ask of the US Navy

Found the Clairvoyant Space Corp as a part of the naval assignment. Annapolis should be one of the Consciousness Center's academic chairs. This covers not just the Look Before You Leap program of involving remote viewers in naval operations but also the political invectives that this type of research must be defended against.

Look currently at the problem the United States has in combating Chinese hackers. They are far outnumbered and are stating so publicly. Your support in this endeavor will add resources to the navy's mission and bring it forward in both technology and thought. Naval aviators will be our first line of defense against hostile ETs and should have nonlocal cognitive support when flying those missions.

Creation of the CSC: In creating a cadre of CSC officers to monitor and protect our aviators on a mission

The Applications of the Science of Reincarnation

with six fighters, there would be six CSC officers remote-viewing the mission. This added bandwidth multiplies the information acquired.

Additionally, the CSC would train all ship's crew in remote viewing and sensing to increase the bandwidth of acquirable information.

To understand how this would work, one must understand what flying cap is in an operational sense. The Airborne Warning and Control System (AWACS) coordinates all activities in the theater of operation, flying above all activity and monitoring all activity. It is called flying cap. The CSC operation would operate above the AWACS and within its theater of operation, providing a new means of data acquisition.

Of application to naval aviators will be the new technology of the ARVS, which can fly in the air, underwater, and into space.

Look Before You Leap should be the motto of this new branch because, at that point, we will have the technology to follow our alien visitors back to their "Mother Ships" or local bases. For clarity to the nonprofessional reader, those bases could be the moon, Mars, Ganymede, or our own South Pole. Additionally, there may be dimensional nonlinearity with some of these locations, causing initial navigation problems and cognitive problems.

When we remote-view alien locations, they know we are there looking, so to hide what they are doing, they stop work until our remote viewers leave. Dr. Courtney Brown has explained this in his book *Cosmic Explorers*.

Without this type of data acquisition, our own aviators are at a disadvantage.

The aviators themselves need to be trained in this with their crews. Consciousness problems will emerge for these crews because once in 4D space, your past lives, past races, and past genders merge into a 4D version of yourself. That is why consciousness training is imperative in our space-bound military.

Here is how I would frame the request to Bill Gates.

Bill,
You understand the need to increase capacity before we use that capacity. This is core to funding consciousness and fusing and understanding 3D and 4D awareness.

Stuart Hameroff's work is putting the how under our observations and the cognitive landscape we see.

Bill, I am not asking you for $20 million, but in terms of MacKenzie's request, I am really asking for $100 million; the $20 million is just to set up the plan. Now why, and what is at stake? Stuart Hameroff needs more computing power than anyone I know because he is trying to count the electrons in the human soul. He would disagree with that rendering, however, being reduced to one sentence. It is what is going on at the Consciousness Center. For your clarity, the electron exists both as a particle and a wave, and everything we talked about in our observations is waveform effects in a particulate environment. Here is the website: https://consciousness.arizona.edu/.

This leads to a new area, a nonlocal space, that shows intelligence in the alien spacecrafts that we cannot deny any longer; they reside just like we do. You are being asked to fund an acceleration in human evolution. The aliens want us to join their federation but say we are not yet ready. Radin's involvement in DNA tracking of nonlocal consciousness genes foreshadows Commander Troi on a flight deck and humans in space, genetically advanced using clustered regularly interspaced short palindromic repeats (CRISPR). I want more than $100 million, Bill, and Stuart badly needs your help.

There is so much more to connect and so much that isn't connected. In the nonmaterialist category, we have stories about interaction with what Brian Weiss calls the "Masters," intelligence forms who we can talk to nonlocally. Stuart Hameroff can measure the body during these interactions, and this same holistic approach can be used to ask pointed questions about this science when we do these measurements. What we are trying to fund involves taking Jim Tucker's children, Brian Weiss's regresses, and Pim van Lommel's NDEs into the lab for comprehensive studies. These forms of nonlocal consciousness can be used militarily.

We are coming to a point where our data management capabilities and our own bodily information to store all the data of the body will be equal.

Our consciousness of birth and death is an upload-download fractal of information management. I would ask you to ask Stuart Hameroff what data management solutions we can provide.

Very truly yours,
TIFACS

You see, Mr. Bigelow, you can get to all of them to present a cohesive plan that solely revolves around human consciousness. Everything they do individually connects to a collective view of human well-being; you cannot have a billionaire class without it. This is in all our interests, and what you are asking them for is cohesive action.

The funding is for the architectural build-out of the science of reincarnation. You personally know this. I know you know this. Once funded, the pieces act cohesively in both 3D and 4D, both forward and backward. For this to work, concepts such as "time has mass" need to be understood as much as consciousness existing without a body.

Now let us revisit rewriting the US Constitution as a square base unit as opposed to the gerrymandered or influenced base unit. Can you see if having the irregular shapes to each unit would slow and foul the system, shit in, shit out?

In terms of global system management, this rework of the US Constitution is a fractal example of producing a common consensus. This is interfered with by gerrymandering. But in terms of global application, a common mind fully informed is a defense against extraterrestrial manipulation.

So constitutional redesign means cooperative space efforts, research and social programs, and defense. Social programs optimizing productivity through the management of resources mean stopping war while saving biodiversity in human populations.

In short, we are aiming for benevolent management of people through wellness metrics, for without it, we cannot propagate into the universe or defend ourselves from it.

We cannot protect ourselves from threats we refuse to recognize. When you speak of theory development in consciousness studies, the standard model of consciousness that universities are being asked to produce will rewrite human history and change our view of religion and gender. Theory development must take this and the urgency to do it into its equation, which means this needs to be funded now—right now.

We have addressed the extraterrestrials.

The transdimensionals are another type of problem.

There are other ways we are being attacked, and we can defend ourselves if we understand how.

In the science of cognition, we are piercing a dimensional wall. The science of reincarnation is becoming a real and valid study in data management. Our collective understanding of this higher dimensional state, which we all can individually access, is not yet explained effectively, and so information on alien presence is best managed. The technology transfer that is occurring is far ahead of our understanding of the scientific and political ramifications.

This explanation of consciousness needs to evolve. When one remote viewer views something, then ten do a better job. Global consciousness is better still regarding understanding extraterrestrial intelligence and extradimensional intelligence and its exploration.

We are remote-viewing the universe. Look to the organization this builds that can be used in the open for intelligence activities. At that point you are establishing a center at the University of Arizona Consciousness Center run by Nobel Prize winner Stuart Hameroff that lays out an intergalactic welcome mat, and we look not with ten remote viewers but ten million. While we do this, we must acknowledge that we can be attacked along the same pathways, and local religious zealots feed on our not teaching globally an idea that limits their influence.

In short, *The Standard Model of Consciousness* is a scientific explanation of their belief system that supports it and expands it to the point that religious differences cease to be meaningful in the larger algorithm. With one book written by the best global scientists, you take down religious zealotry in the next generation and reduce the "noise" that allows the misinformation that manipulates us as a species.

Look at what NASA is asking us, the people, to do. We all have asks, it seems. This is from *The Washington Post*, in an article called "Want to Listen to Space Noise? NASA Wants to Hear from You":[108]

> Note to citizen scientists: NASA wants your help listening in on the universe—specifically, on the sounds of low-frequency waves generated by solar particles colliding with "Earth's magnetic environment."
>
> The goal is for volunteers to assist the space agency's scientists in identifying important features within the cacophony.
>
> Heliophysics Audified: Resonances in Plasmas (HARP) is now open to citizen scientists. HARP uses data from five NASA satellites that launched in 2007 to help study auroras, which occur when

---

[108] Erin Blakemore, "Want to Listen to Space Noise? NASA Wants to Hear from You," *Washington Post*, April 23, 2023, https://www.washingtonpost.com/science/2023/04/23/space-sound-nasa-research/.

solar particles run into Earth's bubblelike magnetic field, or magnetosphere...

Think listening to years' worth of wave patterns is a job for artificial intelligence? Think again. In a news release, HARP team member Martin Archer of Imperial College London says humans are often better at listening than machines.

"The human sense of hearing is an amazing tool," Archer says. "We're essentially trained from birth to recognize patterns and pick out different sound sources. We can innately do some crazy analysis that outperforms even some of our most advanced computer algorithms."

If you would like to hear the sounds of solar wind and more, and to contribute to the project, visit Listen.spacescience.org.

I say this to NASA, and I say it with love. Give me a break. The outline here in this paper is to give you more resources and institutionalize it with a global network centered in Arizona. People cannot remote-view into the cosmos without a clearer and more precise understanding of their own consciousness, ergo, *The Standard Model of Consciousness.* Now you are not just listening like you are asking but actually looking. To do this, you must own up to the shit that went down over the last seventy-five years. Keep your secrets. This is not about prying into your files; it is about trying to save humanity, and

1. you need help;
2. we want to help;
3. the two thousand billionaires who control half the world's wealth can, through this proposal, be actionably accessed and filtered through the best minds we have; and
4. it can work for all of us.

So, there you go, NASA. Best of luck.

For the rest of us, this simple explanation has more than just global defense consequences. It has political and social consequences.

This model of consciousness supports O. W. Holmes's theory of freedom of speech in that repeated and public lies do real societal harm and are, therefore, subject to prosecution under the law. O. W. Holmes and the common mind, and Fox and One America News Network lies, can be successfully scientifically challenged in a court of law.

LGBTQ is now scientifically valid, and opponents such as J. K. Rowling and any school district that shares her views are scientifically invalid, and damages can be assessed in a court of law and can be scientifically challenged in a court of law.

Transdimensional understanding begins with the science of reincarnation, which supports belief in every religion's afterlife as a scientific model and is managed by data management.

Theory Development

If someone asks you what the science of reincarnation says, "It is as simple as uploading and downloading information to the cloud and putting it on another device. All of us are information. That information is written in your DNA and in your body's energy. Once we can process the amount of information in us, and that is not far off, we will have pierced the dimensional wall between us and heaven. Psychics, remote viewers, and mediums do it all the time. This is just the scientific explanation of what they do and the landscape they operate on and one that most people already believe."

The very study of information topology in formalizing this method of data management undercuts scientifically religious nationalism in all places this science goes.

When you evaluate the science of reincarnation, look at the political ramifications of changing belief to understanding. It begins with understanding our own consciousness, which changes the rules by which we make judgments. Change that paradigm, and you change humanity.

This idea gives you great political leverage while at the same time opening the playing field for our intelligence services because it becomes a force multiplier. The intelligence community is sewn into this structure at the very beginning, and while alien-to-human technology is ongoing, what lags is the human community's ability to deal with it. This structure proves heaven, but dynamics change with a 4D understanding. In terms of politics, there are not many religious denominations; they are all part of a common denomination. This leaves religion in place, and family traditions are honored, but there are no longer borders. This can be accomplished by funding the proposals attached.

In a very real way, *The Standard Model of Consciousness* that scientists are being asked to write becomes the scientific Bible of the new age.

In terms of theory development, sociologically, we are better able to defend ourselves with our new cohesiveness. It is harder to insert false information packages. But then the cosmos opens.

By expanding the conduit that the aliens have established through crop circle communication, we connect with an interstellar race that lives on three different planets in their solar system.

With humanoid populations on Mars and our moon, our solar system is similar, and we just are becoming aware of it. This proposal in formalizing the study is to study consciousness, our own development, and how other more advanced societies developed safely and cohesively. The study of our consciousness can be found in our actions.

This paper, at its core, is about opening a second channel to the universe beyond NASA and establishes the right of all humans to communicate with the universe to listen and learn. NASA has formalized the request for us to listen; they cannot stop us from learning. It is not one channel that will open; it will be many. We can help each other. How is that for theory development?

Homer, this chapter is very simple for you. Repeat after me: There is some shit going down.

Chapter 13

Empiricism

The second area where an infusion of funding would be important is experimental research. While the number of researchers engaged in empirical parapsychology has always been rather small, as a group, they have been remarkably persistent, competent, and productive. A half-dozen classes of psi experiments have reached a stage of maturity where methods to replicate effects can be described in straightforward terms, and replications should be encouraged as teaching tools. Part of the funding, then, would create teaching systems to help students replicate experiments known to be successful. This would include, for example, hardware and software for a digital Ganzfeld telepathy system and setups to allow for psychophysiological experiments such as presentiment (unconscious precognition)—the feeling of being stared at from a distance—and brain-to-brain correlations. These teaching systems would be made available at low or no cost for educational institutions and online for no cost.

The rest of the funding would fall into two categories: The first is directed multiyear programs. These would be experimental programs requiring a minimum three-to-five-year effort, with the research team identified and invited by a steering committee of experienced researchers. Longer-term programs, those of up to ten years, would also be considered, depending on the nature of the proposed programs, the track records of the proposers, and the judgment of the steering committee. Second, an international grants program will solicit requests for proposals. These will be offered annually, with a maximum of $100,000 per grant. There is no lack of interest among researchers in studying psi phenomena, but the range of experiments that have been conducted to date has just scratched the surface. With an infusion of funds, the scope of phenomena studied will creatively explode. Besides experimental tests of elementary psychic phenomena, including telepathy, clairvoyance, precognition, and psychokinesis, this effort would significantly expand research on survival-oriented phenomena, including mediumship, channeling, near-death and out-of-body experiences, and reincarnation.

In addition, a worldview survey can be undertaken to find exceptional talents (children and adults) to study, and in future efforts, the experimental program can be integrated with these talented individuals. This effort is estimated to cost about $100 million, with most of the funds going to the long-term and solicited grants program and supporting the positions required to administer and track those grants.

—Dean Radin *the Mathematics of the Science of Reincarnation*[109]

Simply put, empiricism is the idea that all learning comes from experience and observations. The term "empiricism" comes from the Greek word for experience: *emporia*. The theory of empiricism attempts to explain how human beings acquire knowledge and improve their conceptual understanding of the world.

I ask the billionaires, do we have the balls to believe our own eyes?

The proof is in the liars across all spectrums of this information; if I were to do an odds-against-chance calculation of Lazar, Aldrin, MUFON, Greer, Elizondo, Salla, and others all telling the same lie, I would arrive at a certainty despite the denials. This being the case, the Grays have reincarnation technology. Now I ask all billionaires, what are you going to do?

Collective consciousness and global geopolitical cohesion equate to a global intergalactic strategy. Consciousness underpins our entry into the universe.

---

[109] Bob Good, Dean Radin, Stephan A. Schwartz, Titus Rivas, and Cathie Hill, *The Mathematics of the Science of Reincarnation* (Boynton Beach, FL: IASOR, 2020), 381.

This is empiric in its call to humanity as well as to our alien neighbors.

If everyone is lying, Lazar, Hynek, Greer, and so on, then I think a Jedi master is telling me that these are not the droids I am looking for. In short, we are being bamboozled.

While Dean speaks directly about research, the fact is that the derivatives need funding, or the research dies.

Empiricism comes down to what you believe. Did the math here make sense?

So is this presentation true? This document uses discredited sources, certainly ones that have come under heavy criticism—Salla, Brown, and Greer, to name three. But others such as Buzz Aldrin and James Mitchell are trusted sources and provide corroborated facts.

There is one other category that can provide you with the truth: yourself. You look at the news, the videos, and the firsthand accounts, and you can say with certainty that not only are aliens here, but they are among us. Watch the night skies.

This document, then, is not meant to prove they are here but to form a plan of what to do about it. Here is a general letter to send out.

An open letter to all billionaires:

Dear Sir/Madam,
This cover letter is meant to introduce you to an effort funding consciousness science according to the attached plan for $450 million. Additionally,

each of you is being asked to participate as a curator for one part of the overall effort. This cover letter will explain what is being asked of each of you, why you specifically are being asked, where the money is to be spent, and who is involved. It will also explain tangible products that will benefit you individually and personally. This begins with three empirically proven discoveries not yet processed or understood because of their transcendent ramifications.

1. We have scientifically proven a continuity of consciousness after bodily death.
2. We have scientifically proven the existence of alien life and visitors who trans navigate our airspace and oceans and whose presence is current and continual.
3. There are transdimensional beings who live on or frequent planet Earth. Anything humans do to harm the planet harms their environment.

You may disagree with all three premises, but the empirical proof is yours to examine. Twenty million dollars is small enough from each of you, but your participation makes this a cohesive whole in terms of an intelligent, collective, and measured response. Some may not wish to participate; these positions may be taken by other billionaires. It is not just the money you are being asked for; it is your participation/guidance on various aspects of how we, as a species, should respond to this new information.

Each of the recipient organizations, most of which are 501(c)(3)s and specific individuals within, are given "asks," producing a collective intellectual ask overall, with each having part of the whole. Individually many may come to this project reluctantly or not even know of it. But they will come because, like the ask of you, it is the right thing to do.

Given the size of the money and distribution of it in this proposal, some of you are being asked upfront to fund various projects, and some of you are being asked to contact specific individuals to discuss where said contributed money should best be dispersed. Each donor will have specific targets within the overall plan based on their exhibited talents and interests.

In summation, you—the money—are joining the talent—the scientists—to form a team to produce the most effective response humanity can have to the new conditions that are emerging, proving continuity of consciousness after death and aliens frequenting our planet.

The following will introduce you to the other invited donors and the subsequent proposals of how to begin this effort, along with a summary of the purpose for each.

Your first question may well be, what does the afterlife have to do with aliens? Both are fractals of nonlocal consciousness. Consciousness is what we are proposing to study. Telepathy, clairvoyance, out-of-body experiences, and the afterlife are all fractals of our personal connection to cosmic communication, which is a data acquisition multiplier. The business and military applications are transcendent; they are next generation and beyond, and some aspects of this proposal will look to commercialize them.

Truth: These proposals are about funding, where it should be applied, and how it should be applied.

The science of reincarnation is a methodology that operates at all levels of magnitude of consciousness. Data transfer is what it is called for inert data. For consciousness uploads, we reincarnate to a new device, and it is the energy and not the device that is cognizant.

What would you be investing in?

- Bio-transfer technology
- A spaceport in Arizona
- Commerce with the universe
- Changing humanity

These proposals are some of the applications of the science of reincarnation, not just on a personal level but on a societal level. The empirical truth is that if we do not evolve, we will die. To poison the planet is to poison ourselves. The empirical truth is that investing in this plan is investing in your future.

Homer, if you have read this far in this book, even at your level of understanding, you know what the truth is. That is empiricism.

Homer, there would be four people I would like to see sit on the empiricism committee: Stuart Hameroff, Dean Radin, Stephan Schwartz, and Joe McMoneagle.

Chapter 14

Army Futures Command: Military Analysis of the Data and Resultant Strategy

If the previous chapter, "Empiricism," was about truth, then who would handle the overall implementation and coordination of such a plan? An organization large enough, with a skill set in understanding the situation and able to implement programs to benefit all of mankind? In short, who manages the defense of our world?

While many would say the United States, that would not be specific enough. It is not just the military but a specific branch of the military, the United States Army Futures Command. The following is what Wikipedia has to say about them. What they say about themselves is that they are always preparing to fight a war that will occur thirty years into the future. They are preparing our military now to fight that fight then.

That's wonderful, but they/we are overmatched now, and we will be overmatched then. This TIFACS plan is about bringing them resources they will need by harnessing half the world's wealth and all the NGO talent to be put to use in various aspects of meeting the challenges of joining a galactic neighborhood we have been describing.

I want to remind you, my reader, what Col. Karl Nell said at Salt. "There is no plan." So, this TIFACS plan is not just to fund researchers and to make billionaires money or make them altruists; it's to save the fucking ass of the planet. Even Homer Simpson would understand this.

What the US Army Futures Command must realize is you cannot defeat radical ideology on the battlefield. That means there is an ideological battle going on here on Earth now, and fighting the battle thirty years in the future will need us united as a planet and a race, so their ideological battle is here, now.

So let's let Wikipedia introduce them.

> The United States Army Futures Command (AFC) is a United States Army command that runs modernization projects. It is headquartered in Austin, Texas.
>
> The AFC began initial operations on 1 July 2018. It was created as a peer of Forces Command (FORSCOM), Training and Doctrine Command (TRADOC), and Army Materiel Command (AMC). While the other commands focus on readiness to "fight tonight," AFC aims to improve future readiness for competition with near-peers. The AFC commander functions as the Army's chief modernization investment officer.
>
> Army Futures Command was established by Secretary of the Army Mark Esper to improve

Army Futures Command: Military Analysis of the Data and Resultant Strategy

Army acquisition by creating better requirements and reducing the time to develop a system to meet them. Between 1995 and 2009, the Army spent $32 billion on programs such as the Future Combat System that were later cancelled with no harvestable content. As of 2021, the Army had not fielded a new combat system in decades.

A fundamental strategy was formulated, involving simultaneous integrated operations across domains. This strategy involves pushing adversaries to standoff, by presenting them with multiple simultaneous dilemmas. A goal is that by 2028, the ability to project rapid, responsive power across domains will have become apparent to potential adversaries.

2019

AFC declared its full operational capability in July 2019, after an initial one-year period. The FY2020 military budget allocated $30 billion for the top six modernization priorities over next five years. The $30 billion came from $8 billion in cost avoidance and $22 billion in terminations. More than 30 projects were envisioned to become the materiel basis needed for overmatching any potential competitors in the "continuum of conflict" over the next ten years in multi-domain operations (MDO).

From an initial 12 people at its headquarters in 2018, AFC grew to more than 17,000 people across 25 states and 15 countries in 2019. Research facilities and personnel (including ARCIC and RDECOM) moved from other commands and parts of the Army such as the United States Army Research Laboratory.

Organization

The commanding general is assisted by three deputy commanders.

- The Futures and Concepts Center: The first commander was AFC deputy commanding general General Eric J. Wesley, and it was led in 2021 by Lieutenant General Scott McKean. The center operates along four lines of effort: science and technology, experiments, concepts development, and requirements development.[110]

You cannot deploy new concepts without discarding old falsehoods. This must be done for common self-protection. At this point, Army Futures Command, in meeting the external threats and opportunities that Earth

---

[110] "United States Army Futures Command," Wikipedia, https://en.wikipedia.org/wiki/United_States_Army_Futures_Command.

Army Futures Command: Military Analysis of the Data and Resultant Strategy

must face, must act as the agent for the globe and not just America, until an organization with a clearer vision can take its place.

This cannot be done unless adversaries become allies and we stop killing each other. Bear in mind we have proof that these alien groups have been around for thousands of years longer than us, and we have already explained using the Hubble telescope metric how many there could be.

Given they could have had us at any moment means the correct action going forward will keep us safe. We must learn their telepathic language and develop our own natural ability.

Additionally, these aliens are transdimensional. The technology transfer has been going on for some time but unacknowledged by our government. **There will be a moment, however, when it will not be sustainable without an upgrade to the common man's understanding of his place in the universe. The things upon which we fight about—race, gender, and religion—are impediments to a world of peace, a threat to our galactic neighbors, and meaningless in the 4D reality we inhabit.**

Secrecy among branches of government is detrimental to the entire organization. While the government resists disclosure, it should allow the creation of new informed communication channels that don't threaten existing secret programs.

The way the US Army Futures Command can change this now is simple: Publicly do the research and explain it to the common man globally. Hindu and Sikh, Christan, Jew, and Muslim are all the same. Teach the science of reincarnation globally and in a generation, thirty years, you will have a unified global understanding that will gradually replace the existing structure of regional confrontation.

It will leave in place our heritage and traditions, **but change the idea that one is better, when it is better that they are one.**

With that single act and the organizational resources that TIFACS's plan lays out, the AFC will have the resources to meet the challenges that the future holds.

In the meantime, TIFACS, which will be housed at the Monroe Institute, will design teaching platforms to explain the gestalt of the situation and work for funding to assist the Monroe Institute in joining with the Consciousness Center at the University of Arizona, where training programs and teaching manuals will be printed in all languages and distributed through the twenty chairs of parapsychology.

That teaching hub will include the following schools based on the twenty chairs of consciousness study so the Future Command builds out a research and consciousness network outside of its budget using not just the money but the talent of the scientists and billionaires. There is your plan.

Army Futures Command: Military Analysis of the Data and Resultant Strategy

- **Combat Development**: Helps AFC commander to assess and integrate the future operational environment, emerging threats, and technologies to develop and deliver concepts, requirements, and future force designs.

Emerging threats can simply be explained by the ever-growing list of alien species present in our skies, on our planet, and in other dimensions we occupy.

- **Acquisition and Systems** (founded as Combat Systems in 2018):
- Gen. Robert Abrams has tasked III Corps with providing soldier feedback for the Next Generation Combat Vehicles CFT, XVIII Corps for the soldier feedback on the soldier lethality CFT, the Network CFT, as well as the Synthetic Training CFT, and I Corps for the Long Range Precision Fires CFT.
- Combat Systems refines, engineers, and produces the developed solutions from Combat Development.

Emerging opportunities for technology transfer are already in place to be enhanced by better governance and management.

This plan is designed to acquire the cadre of our current enemies and synchronize our goals in dealing with the aliens in our time and space and those with transdimensional capabilities. To address the transdimensional capabilities, we have proposed the Clairvoyant Space Corp.

Do you think, General, that the intelligence services would love to have access to the proposed network? Caution: This network should disseminate true science only, not propaganda. The truth is the propaganda, news, message, and so on. At the same time, you benefit the people receiving the message and who join the organization because you are telling the truth.

We have proposed a training program developed and administered by the Monroe Institute, the Consciousness Center at the University of Arizona.

Forward-leaning investments in programs such as this need a mandatory national service to develop more connective tissue. Young Americans can meet people from other ethnic groups and other sexual orientations and realize they can build something great in the agency of others and not see people as Republican or Democrat or transgender or gay and see each other as Americans and humans.

The increased modernization and additional capabilities will come from the symmetry, which is why **TIFACS** will be housed at the Monroe Institute. Let us look at the Monroe Institute's vision of itself.

> Monroe's "Vision for the Future" represents their goal of reaching 1 percent of the world's

Army Futures Command: Military Analysis of the Data and Resultant Strategy

population with a direct experience of expanded awareness so they know they are more than their physical bodies.

TIFACS has chosen Monroe because of Monroe's long-standing relationship with the CIA and the US Army.

Homer, nod, nod, wink, wink. Homer will get the movie reference.

Now to be clear, TIFACS to Monroe is a force multiplier. TIFACS calls for a $6 million contribution to Monroe, twice its annual budget, to support achieving Monroe's goal more quickly.

So let us look hard at Monroe's goal of reaching 1 percent of the world's population with direct experience of expanded awareness.

We have spoken about systems already, for instance, the upload-download system, the systemic application of math analysis, and systems development using odds-against-chance calculations to search for fractal patterns that indicate our true reality.

But what about systems that do not work, such as global governance? That system is governed by confrontation and regional self-interest. How does this system change, and what is the tipping point?

If the context changes, then the individual data points begin to reorient, and at a certain point, the entire system undergoes radical realignment to the new paradigm.

If I could speak to General Rainy of the US Futures

The Applications of the Science of Reincarnation

Command, I would say, General, I listened to Karl Nell's interview at Salt and the lack of a plan and the potential for societal disruption. I offer the science of reincarnation as a bridge between nondisclosure and disclosure and a useful weapon in the United States arsenal in fighting on so many fronts. Please allow me to explain.

The premise is that the work at UVA by Jim Tucker (children who remember prior lives), in Miami by Brian Weiss (past-life regression), and at UNLV (near-death experiences) by Raymond Moody are fractal proof that this is our reality. Access to this new reality is through the Gateway Process. There it is called Locale II. That is the science of reincarnation.

What happens if the US government comes out to fund the science of reincarnation for $450 million and says where there is enough smoke, there's fire, and gives our friends and enemies the statement that they believe in research here?

There is no societal disruption in saying to all religions that their view of the afterlife is real and that all species with an electromagnetic imprint exist there.

Once you create a 4D reality recognized by the US government, truth becomes self-evident as both your cadre and your enemies explore the data and you eliminate the replay of the rise of Hitler using the same slogans from a century before.

If the battle thirty years out is on Earth, humanity is lost; if humanity is united, we may have a chance. Articulate

Army Futures Command: Military Analysis of the Data and Resultant Strategy

that publicly in conjunction with the announcement of the science of reincarnation funding. Prepare that battlefield by subverting the cadre of your enemies now and acquire your enemy's resources for that battle.

What does this mean for the leaders of autocratic countries who find the United States validates this research in this way? What does it mean to Xi and the Uyghurs if retribution can come after death and in front of his ancestors? What would it mean to Putin to find out we are doing serious science on life after death and the consequences of actions?

What does this mean for any Islamist extremist or conservative evangelical when their religion is both validated and democratized? Perhaps this thought is antithetical to your position to prepare for future wars; this is designed to stop the war, a requirement from the aliens who will allow us some technology, but before we can transition to a galactic species, we must stop killing each other.

Acknowledging reincarnation inevitably leads to exopolitics. For humanity to address this disclosure must occur in a way that is relatable, the transition to a united and healthy world to interact with the larger universe. Humankind needs a transitionary tract that leads to a cessation of war and a democratic unification to benevolence or suffer the fate of the resistance or races much older than ours.

The strategic initiative to establish the science of reincarnation is an actionable tactical act that will bring our enemies' resources to our use and their benefit. This is a win-win-win strategy.

Finally instituting such a program would indicate to the alien groups observing us that humanity is moving to cultural unity and an early understanding of dimensionality and the environment around us.

While there are credible reports of technology transfer from off-world intelligence, humanity cannot advance unless it stops fighting itself. This event benefits the aliens who would have an evolving, more stable, and peaceful human population. This is resource-positive and nonthreatening to exo-actors and is an attempt to bring peace, stability, and clarity to our existing reality.

No longer tribes with different gods, but a common consciousness emerging, and a deeper understanding of our place in the universe and its dimensions we occupy.

Perhaps I could trade a speedup of technology transfer to preserve this planet in exchange for a program to stabilize human ideology.

Publish this and the Gateway Process together and every intelligence organization will be running to see what you are doing. You have just disrupted the religious beliefs of their cadre.

There are no losers in this proposal; you free the next generation from past ideology.

At the end of chapter 5, "Humanity's Current Situation," we said whoever is going to deal with the problems we present is going to need a bigger boat. *Jaws* (1975) is the movie reference.

So here is how to build a bigger boat.

Wikipedia said, "Between 1995 and 2009, the Army spent $32 billion on programs such as the Future Combat System that were later cancelled with no harvestable content. As of 2021, the Army had not fielded a new combat system in decades."[111] What I would suggest is that you suggest, off the record, to any one of the aerospace subcontractors or weapons contractors you might meet that you would like a donation to TIFACS sent to the Monroe Institute as a founding donation of $20 million. They sit on the board; Monroe sits on the board. That is two.

Monroe will act as the fiduciary and found the 501(c)(3) for TIFACS, and immediately TIFACS will donate $3 million to Monroe, effectively doubling their annual budget. With that donation, Monroe will reach out to the University of Arizona Consciousness Center to begin discussions about the Gateway Process being taught at the University of Arizona as a three-credit course. This includes courses on the science of reincarnation using the textbooks of *The Science of Reincarnation, The Mathematics of the Science of Reincarnation,* and this one, *The Applications of the Science of Reincarnation,* along with Kroth's book on crop circles, *Messages from the Gods,* and implementing

---

[111] "United States Army Futures Command," Wikipedia, https://en.wikipedia.org/wiki/United_States_Army_Futures_Command.

this program as guideposts on our journey to global consciousness and understanding our individual and collective place in the universe.

I would briefly point out that for all our sophisticated equipment, it was an amateur astronomer who detected the comet about to hit Jupiter.

That leaves $2 million for TIFACS. Monroe will then provide one full-time administrative assistant to TIFACS at Monroe's offices, and that person will then begin to solicit a $20 million donation to the Consciousness Center to accept and accredit the Monroe Gateway course, and any others they agree to. This administrative assistant will have a lot on their plate organizing all the Gateway-related threads, to include Reddit, Instagram, TikTok, and so forth, beginning the search for the first executive director and general managerial work.

The Consciousness Center gets a seat on the board for accepting as well as the money, all $9 million. Now there are three.

There may be other military contractors that may begin to see value in this, and there is early danger of corporate control, but it must be understood by all that a lack of control may benefit all. A contributor to this, you can look to maybe an aerospace contractor who understands the possibility of technology transfer when planes get flown by thought. Bigelow Aerospace would and should get a seat with a $20 million donation.

Army Futures Command: Military Analysis of the Data and Resultant Strategy

The lines of research must operate outside conventional understanding so the freedom accorded the scientists in this endeavor must, as an imperative, be protected. To that end and to protect that initiative, the following are given board seats. Five science seats are to be determined by election in an open session at their annual meeting. Scientists select those five from themselves at their annual meeting. The Consciousness Center gets the chairman and five elected scientists. TIFACS gets a seat. That is nine seats.

This introduction to the topic is presented to the community of consciousness researchers, who will no doubt take issue with the distribution of funds I suggest in this plan. I say this to all of you. This is the beginning, and you need to organize what you need to be funded at the grassroots level. Articulate your needs and the required funding and connect it to the larger study of consciousness. Then support the creation of TIFACS by posting your funding request on our website. Only by establishing the structure to acquire new and meaningful sources of revenue for research can we go forward optimally. For every ten projects funded in this way, nine will fail, but not all failures will be failures; they will narrow the path to success.

Once that is done, what do you think TIFACS could sell/rent a seat on its board for? This may seem like a strange question here, but you are fusing consciousness research, a priority, with defensive preparedness, a necessity, with simply sound business, an expediency. My point is that this becomes self-funding after an initial push, and we go not just go galactic but go transdimensional. Every copyright is part of the branding.

If you are a next-generation Islamic radical and you are free to go through the Gateway Process, are you still an Islamic radical when you meet the Grays, Tall Whites, or Reptilians? That person will still be Islamic because that is his heritage, but radical? I think not. How do they relate to their peers because they can see how religion is viewed from the afterlife? Just give them Mellen-Thomas Benedict's account and tell them the truth, that this is what happens, and let them sort it out.

This will draw the best and brightest from your current enemies so they can join with us to protect all of us in the future.

TIFACS will also set up a combined education program including Monroe and Arizona literature for world distribution. In short, the architectural design for a global network of trained psychonauts whose mission is to Go Galactic©, and they are being recruited globally for this endeavor.

This simple act of preparing for the future now has political ramifications in the present by virtue of the fact you are building your own cadre inside the cadre of your current enemies, and the transparency of this plan is its danger, the same danger truth has to lies.

You create within your organization a unit called the

Army Futures Command: Military Analysis of the Data and Resultant Strategy

Clairvoyant Space Corp©, whose mission is to remote-view places and events to keep our troops safe in field operations.

As I probe the possibilities here, there is one I find interesting: voluntary continued service after death. If my consciousness continues after my death, then so would my social imperatives, believing in freedom and justice for all. The question I would pose to you is, how could I help you after my own death? How could your entire staff? The world I grew up in no longer exists. The world I am going to I want to be better.

A total of $32 billion were spent with no harvestable content. I want $450 million to start. Four hundred fifty million divided by thirty-two billion is 0.015. A percent and a half of what you have spent with no harvestable content. For that number, you get a global network of psychonauts who can guide you in acquiring weapons systems for your ever-changing needs but at the same time shake the cultural identity of every enemy of freedom, because in the end, we are all equal in death.

You not only have a trigger to this plan; it is also a whisper over coffee to a contractor whom you have spent a lot of money with and gotten nothing in return. It is also plausible deniability because this plan is published with a book called *The Applications of the Science of Reincarnation*, so any billionaire who might feel so inclined could pick it up and run with it all by himself for a variety of reasons.

Once Bigelow puts in his $20 million, $3 million goes

The Applications of the Science of Reincarnation

to Monroe and the search for the COO of TIFACS begins, and TIFACS has $17 million left.

TIFACS then starts the negotiation with the University of Arizona, which includes the Monroe expansion. Once that starts, General, I would like you to lean on contractors for a fast $100 million.

This is not a shakedown but an investment, and time is of the essence. They are investing in new technologies that respond to thought. You just harnessed the best cognitive scientists in the world, and it did not cost you a dime.

The contractors have a new resource. Intelligence has a new information collection network that is multitiered, and the scientists get support they otherwise would not have. Win-win-win.

There is no finer group of consciousness scientists in the world than connected to the Consciousness Center at the University of Arizona. By Bigelow BICS sending this proposal to two thousand billionaires, you engage half the world's wealth and most of its talent.

So here is a short version we can call frequently asked questions (FAQs).

Why Bigelow to head the NGO, TIFACS?

He is the most knowledgeable of all the billionaires and runs both Bigelow Aerospace and the Bigelow Institute for Consciousness Studies (BICS). That organization just ran an essay contest with a first prize of $1.5 million with the question, "Prove continued consciousness

Army Futures Command: Military Analysis of the Data and Resultant Strategy

after permanent bodily death." In doing this he created a panel of global consciousness judges for the contest and had entries from consciousness researchers globally.

AFC should contact Mr. Bigelow to discuss interest and structure. Deliver $100 million to Bigelow, who is the fiduciary and has skin in the game, simply by asking for the support of your contractor group. The entry fee is $20 million delivered to TIFACS c/o the Monroe Institute.

Why should AFC support this, and what could they possibly get? What are some things that could occur if this plan is implemented?

To Military Futures Command, I would ask if you could use this very vehicle to ask all Russian billionaires to join your billionaires in this effort. The only ticket to entry is adopt TIFACS intergalactic policies because they resonate off-world or when Putin falls and the oligarchs lose it all. Again, it would be a $20 million entry.

Would a global network of chairs of parapsychology be of benefit in identifying races communicating with earth right now?

AFC introduces a policy that if you go through the Gateway Process, you get a psychonauts badge. Then challenge the Imams of southern Egypt to go through the Gateway Process or their cadre. You are telling your enemy the truth for him to see for himself; you both face the same problems because of this new information.

This vehicle turns Bernays's information system toward

human improvements. The rules are still the same; the game has changed for the better. If races, which are demonstrably here and five thousand years more advanced than us, see an emerging galactic partner, it provides us a safety net to grow, which is better than collapsing from within.

There is evidence that advanced civilizations do not want to interfere with us unless we ask for help. We must earn that help by being someone, a race that merits it by the fruits of our own endeavors. You are not just fighting the war here but there as well.

The best way I can explain this is that a strategy that works in one set of circumstances may be unsuccessful in another.

No new harvestable systems applying 4D groupthink to 3D and in thirty years 4D issues.

Now look at the consciousness proposal in chapter 9 and the databases of international students who pass through those institutions. They are your cadre in foreign markets. They are your global RNGs.

TIFACS at the Monroe Institute and the Consciousness Center export and support the Monroe method globally so people have direct evidence of this themselves. To start you unify 1 percent of the world through Gateway.

You see, sir, the science of reincarnation is not just individual; its societal and collective consciousness can be increased by being taught.

The smart galactic strategy is Earth/humanity

becoming a galactic partner, cohesive in thought and intent. This will attract like-minded races who will incorporate earth in a larger safety net and resource stream.

Change in sexual perception and acceptance by Gen X (the new one) will be a generational shift. The science explaining sexuality in 4D terms and the new generation realizing we are more than our bodies removes friction from the political operating system and removes a weapon in global despots' arsenal.

Sourcing information from the TIFACS organization of experiencers across the world as sources of information, as well as channels of distribution of information through the twenty-one Russian Republics.

Using a military strategy of deploying the billionaires' resources, of not just money but also talent.

Build a global network of information acquisition.

Intelligence is about acquiring information; the Army Futures Command is about deploying it. Write a standard model of consciousness and teach it globally; you cannot tell falsehood when it is truth that will protect you.

Goal: diversity and strength as a species, LGBTQ diversity, DNA diversity. Before you can fix war, you must fix humans and explain to them they are more than their physical bodies.

Futures Command must act and organize a united response to these conditions. You are fighting the battle of the future thirty years out right now.

Army Futures Command, this is a call to action, even if it is just a whisper.

Homer, what I am telling the general is that for $450 million, I can deliver new technology and install it in the middle of the cadre of our current enemies

Homer, there is some cool stuff on the IASOR.org website. Want to join the Clairvoyant Space Corp?

How cool is it to ask Mr. Bigelow to manage a fund of even twenty-three billionaires dedicated to this effort? That is twenty-three times $20 million, or $460 million; with just twenty-three billionaires, the whole plan gets funded.

Homer, the commanding general of the AFC needs to understand that secrecy between branches of government can be detrimental to the entire organization. He needs to be proactive in opening channels.

Chapter 15

TIFACS Policy Positions

What should we do if these are the conditions of our reality? What should our policies be? This may be a difficult chapter to read for some because it contradicts your worldview.

Our policy follows the science and the conditions.

We can explain the dynamic gestalt of our situation:

1. There are aliens with many differing agendas currently coexisting with us in our time-space on our planet and moon.
2. There are transdimensionals. Transdimensional technology is available to extraterrestrials.
3. Continuity of consciousness: Humans have a continuity of consciousness that transcends our own deaths. We are undergoing an evolutionary transition in our awareness of our reality.
4. Human consciousness has a common belief system that life continues after death. It is a commonality among all the earth religions. On this point religions are all the same. Just dress and custom are different to make people be more submissive and

controllable. Continuity of consciousness is a fact for us all.
5. Science is pushing governance to react to the new emerging conditions. An example of this is the push for disclosure.
Health can be infected to harm the body, and infectious viral ideas must be laid to rest by courageous scientists before they infect the mind. Ideas antithetical to logic must be addressed even when they contradict religious belief.
6. Our government is outmatched. Government, just who do you think is getting into those ships you have? Is the resistance to disclosure for societal control?

TIFACS Policy Positions

Our policies must change to protect us all, even those that hate us. Our policy must follow the science and the conditions.

TIFACS policy positions are derived from the underlying consciousness science. The science committee determines the policy positions with a subcommittee to present countervailing views.

These positions are based on sound science as outlined previously. It is up to all of us to teach this. This transcendent logic-driven operation has political ramifications in LGBTQ, religious, and racial legislation. This is a political position based on science, knowledge, and truth.

Let us look at the next election cycle to see how this acknowledgment of our place in the universe affects the evangelicals, the conservatives, our global partners, and enemies, including radical fundamentalist enemies.

AI: There is no getting around the coming sentience of artificial intelligence.

Aliens: Acknowledging their existence near us and the means of telepathic communication they use should change our perspective about fighting each other. Divide and conquer is a strategy best used if the target is unaware of its use. We, as a race, need to be more open about what is happening and aware of its dangers and benefits.

Abortion: The narrative in this is that consciousness transcends death. If consciousness is seeking to be born, then aborting a fetus would only mean that that consciousness would seek a birth elsewhere. Why would consciousness seek to be born in a child with a congenital disease that would die in six months? Or be aborted? In the data are examples of older souls choosing to take this route because they were helping younger souls experience love and the pain of loss.

The Applications of the Science of Reincarnation

Education blocks false narratives; do not be bullied into believing lies. Educate the liars with an ideological vaccine. Rather than disinformation, we will provide real information to teach and monitor the comings and goings of beings too small for our current technology to detect. Yes, they are here, and they have cloaking devices.

Civics must be taught. Create connective tissue between cultures—mandatory National and Global Service for cleaning and restoring the Earth. Grants will be available.

Fundamentalists: Fundamentalists in all religions must change their policies to adapt or diminish. Their children will save them because they, more than their parents, understand the fundamental shift in our reality, the data processing, and the fact that fundamentalists should want to save people of color, transgenders, and the complete LGBTQ spectrum because it runs through them and their second bodies as we explore consciousness from life to life. It is not them and us; it is only us. The Taliban has this problem.

We are faced with new perspectives from two separate but common events. Both that aliens in our present interact with us and our proof that the afterlife exists but from a universal rather than an individual perspective. All views are incorporated into the whole, making it strong and healthy. What follows, then, are conclusions drawn that affect us all.

Monroe is to organize experiencers and their data and plan their requests to go out through the network.

The perspective is that we not only need to protect biodiversity on this planet but among ourselves as well.

Good intent is protecting that diversity, and that includes protecting fundamentalists. Diversity and inclusion provide allies and protectors for yourself and others. This policy is universal and in all cultures.

Exopolitics is a mandatory course of information and a new perspective for humanity. Exopolitical action means making us healthy and strong, all people, everywhere.

If you are a fundamentalist, degrading any part of that is lessening your own protective layers, both in the material world and in the energy world where you and the aliens both operate.

Gender equality: Since the data shows that young souls tend to incarnate predominantly in one gender for the first ten lifetimes, and thereafter the distribution between genders is fifty-fifty, if you are going to live lives as both genders, then gender equality should be a given. The consciousness of gender breaks into many pieces.

Once we have established life-death-life patterns of reincarnation and the transitive nature of sexuality, we realize this topic is not centered on science but as a way to divide us and becomes a weapon in the hands of those who would rather see us fight each other than understand and support each other. Ingo Swann is right about societal manipulation from exopolitical forces. We no longer question whether aliens exist but rather must deal with each of their political agendas.

Gender identity: "Consciousness is fundamental; matter is derivative," says Plank. To everyone who says there are only two genders, they are looking at the derivation, not the fundamental issue that unites us all, though. So J. K. Rowling and Elon Musk need education on this emerging science. Mr. Musk must cross these boundaries when connecting mind to matter. That can be done on the biological level; it is harmonizing the mind frequencies where he will encounter conditions similar to those he sees in his daughter. I mean no disrespect here; Mr. Musk has misidentified biologic markers and should reassess his position considering new information.

So here is what I am going to ask of Mr. Musk.

Elon,
I would like you to write a check for $20 million and give it all to Bigelow to put into TIFACS. It is tax-deductible. That $20 million will be applied in two ways. The first is a feasibility study to determine what resources that come through the

Consciousness Center can be brought to bear on your specific problems.

You see, Elon, the woke mind disease that you claim your daughter has is an exceptional talent of being able to see the 4D reality and being brave enough to have her internal imperatives align with the 3D event she is experiencing. You should be very proud of her and learn from her.

LGBTQI+ issues: If consciousness is electromagnetic and the body is a disposable vehicle, then that cycle of sexual change carries from life to life and explains why people feel they are in the wrong body or have sexual targeting issues from a past life. They are in a transitive state of consciousness.

LGBTQ is now scientifically valid, and opponents such as J. K. Rowling and any school district that shares her views are scientifically invalid, and damages can be assessed in a court of law and can be scientifically challenged in a court of law. Transdimensional understanding begins with the science of reincarnation.

Now an interesting thing happens when you make this contribution. Yes, you benefit one of your core groups by educating everyone to the fact that LGBTQI+ is just different expressions of sexuality, and we all travel a circular path over many lifetimes. So these idiots who say it is only genitals that make the gender are exposed as people who don't know who they are, who have not been exposed to consciousness studies, and who have many surprises in store for them.

But this goes well beyond that one issue. To every greedy billionaire who continues to amass money, consciousness studies just opened huge alien markets if those markets want us. If they do not like us killing each other, then to acquire access to those alien markets, we have to change our behavior. What are the rules? There are ways to ask this question and get answers, but they travel on specific channels. I need money and organization to travel and study on those channels because we are studying consciousness.

So, Mr. Musk, please give us your support. Send your check for $20 million TIFACS c/o the Monroe Institute in Virginia.

Health care: Create a global health care system rather than a disease care profit system. The US Army/VA program for whole health should be expanded globally. Teach the world mindful breathing and the Gateway Process.

White supremacy: I will not address the hypocrisy. The solution lies in global education, free at all levels, of real and true science and fact. This change will be generational. The individual reading this may want to pause because the world you are creating here is the one you will return to after your death and before your rebirth.

Would it make you feel better if I say this is God's will? White supremacy is a myth and a teaching tool to unschooled minds, the group known as useful idiots. It is a lie. It appeals to both traits, fear and greed. The lie of religion is that they are the one true religion. We have scientifically shown that this is not true. We have also shown that continuing to believe prejudicial myths hurts us all. We cannot look outward to alien species much older than us and continue to kill each other over such falsehoods. This is a failed ideology, and the science of reincarnation is an ideological vaccine. We must teach and protect the Christo fascists. This educates their children.

Instead of fighting with you, I am fighting for you, trying to protect you. It's all of us, not just some of us, and certainly not just you.

The Taliban are facing drought situations due to climate change. God helps those who help themselves. It is not enough to say if God wants things changed, it will be so; it's up to you to make that change.

Girls and women must be educated because their men will need their support as the world changes.

Religious laws to remove apostasy, blasphemy, and celibacy. Religious laws are rewritten to preserve history, address reality, honor our past, and acknowledge our future. The fetus is not the problem if the child is condemned to a life of poverty.

This must be printed in the twenty-one Russian republics so they have a scientific base upon which to refute policy from the Kremlin and to be signatories to a common constitution and preserve diversity.

Russia: To the Russian leaders (and there will be Russian leaders after Putin), you must stand down and open your country. We will try to help. There is no loss of power here, just an improvement in your process. It would be best if you imported planks from the common constitution into your constitution, as it is in your best interest and those of your people to standardize on the safest and best way to protect humanity. This also protects you as humanity jointly addresses the presence of aliens who are much more advanced than us living here.

Russia must adopt policy planks of the TIFACS Common Constitution as a working tacit global agreement with every government and billionaire to change Earth society to an interplanetary understanding that we are dealing with races thousands of years more advanced than ourselves.

US national defense: By any reckoning, the US government has had dealings with aliens since Roswell and long before. Global defense moves from a reductionist strategy to an additive strategy.

The strategy of funding consciousness science is to recognize and promote the fact that it benefits any opponent you might have; it is a classic win-win, and if three parties are involved, it is a win-win-win. That includes alien societies. This study promotes open markets, global peace, new technologies, and new jobs, and all stem from this funding.

There can be no advancement societally without global education, and education about consciousness is a prequalifier for entry to what we will refer to as space-side activities rather than earthside activities.

In conclusion, this proposal addresses initiatives in our geopolitical interests, ideologically invalidating religious nationalism in the interest of global cohesion. Our reality has changed, and teaching this as a strategy weaponizes this information globally.

Our best defense against an alliance of aliens who are far beyond us and who wish us no harm is to tread lightly and build a structure that reflects theirs. The duplicity of any means on our part is subject to dissection in a world where remote viewing is a fact. Our using an operating system of scientific truth is again weaponizing fact against fiction for our own global safety.

A review of this through the lens of enterprise risk management makes this study a risk we must take if alien groups are predatory. Or if there are competing groups of aliens, opening a dialogue would be for human benefit.

What Ingo says is that aliens are living on the moon and manipulating humanity so as not to develop our sense of telepathy. In a higher sense, they are manipulating our collective consciousness. Again, this may not be one race of aliens but many, and there may be competing groups.

Ingo's point is that we, as a species, are unwitting targets of this strategy, and your money that TIFACS is asking for goes to address this problem in a real sense. Your money goes to global education, which will collectively protect us and our acknowledgment of our place in the universe among a population of space-traveling humanoids who have genetically altered us. Your contribution goes to acknowledging that reality. That acknowledgment from you in terms of the dollars needed to address this problem changes the world.

You exist in a 4D reality and will cycle back. The object here is to improve the place for your return and enhance your experience by improving both you and the location.

TIFACS supports the idea of a common constitution. Not rewriting every country's constitution but writing one general constitution with demountable planks that benefit everyone. This should begin to standardize so many human endeavors. A clean planet, a common and united response to alien presences, color, gender, and social equality, and all this based on the science that affects us all. This endeavor and these policies benefit and protect each one of us.

All of us must take responsibility, both personal and political, as Kirsten Gillibrand is, but she does not understand how it can be explained to her or others. All would benefit if there were a plan to deal with disclosure. But there isn't, so politicians should proceed as follows.

Mrs. Kirsten Gillibrand,

You are legislating to force a disclosure event on information that is being trickle-truthed and managed. Some things are best managed outside politics and government.

You need to do your own due diligence; this is a guide to expanding your understanding. The arrival of these documents to you is as unusual as the topic. Failure to get this right is the stuff your nightmares are made of. This is sensitive information; treat it as such.

In the science of cognition, we are piercing a dimensional wall. The science of reincarnation is becoming a real and valid study in data management. Our collective understanding of this higher dimensional state, which we all can individually access, is not yet explained effectively and so information on alien presence is best managed. The technology transfer that is occurring is far ahead of our understanding of the scientific and political ramifications.

When one remote viewer views something, ten do a better job. We are remote viewing the universe. Look to the organization this builds that can be used in the open for intelligence activities. At that point, you are establishing a center at the University of Arizona Consciousness Center run by Nobel Prize winner Stuart Hameroff that lays out an intergalactic welcome mat.

What does this mean to you, and why would you consider this both actionable and imperative? This simple explanation has political consequences.

This model of consciousness supports the O. W. Holmes theory of freedom of speech in that repeated and public lies do real societal harm and are therefore subject to prosecution under the law. Therefore, Fox and OAN's lies can be scientifically challenged in a court of law successfully if the decision is based on science and the law.

If someone asks you what the science of reincarnation says, "It is as simple as uploading and downloading information to the cloud and putting it on another device. All of us are information. That information is written in your DNA. Once we can process the amount of information in us, and that is not far off, we will have pierced the dimensional wall between us and heaven. Psychics, remote viewers, and mediums do it all the time. This is just the scientific explanation of what they do and the landscape they operate on and one that most people already believe."

The very study of information topology in formalizing this method of data management scientifically undercuts religious nationalism in all places this science goes.

When you evaluate the science of reincarnation, look to the political ramifications of changing belief to understanding.

The tradition of the cabinet dates to the beginnings of the presidency itself. Established in Article II, Section 2, of the Constitution, the cabinet's role is to advise the president on any subject he may require relating to the duties of each member's respective office.

The cabinet includes the vice president and the heads of fifteen executive departments—the Secretaries of Agriculture, Commerce, Defense, Education, Energy, Health and Human Services, Homeland Security, Housing and Urban Development, Interior, Labor, State, Transportation, Treasury, and Veterans Affairs, as well as the Attorney General.

How can they advise when they have no idea what is going on?

To do anything else makes us more vulnerable than we already are.

So, create a cabinet position that addresses exopolicy. In short, a look at off-planet politics, opportunities, and threats.

The first thing this cabinet post will do is commission a public roundtable whose mandate will be to provide as complete a picture as possible from the public sector of both 4D opportunity and the state of 3D policies and contact.

Panel members shall consist of Laurance Rockefeller, Lue Elizondo, Robert Bigelow, Stuart Hameroff, Jennifer Pritzker, and Jeffrey Mishlove, the current CEO of the Monroe Institute. Also, David Greer, Joe McMoneagle, and the head of MUFON, the head of research at the Skunk Works and NASA. It will provide money for the study of crop circles per Jerry Kroth's request of $15 million. They will paint this picture without asking our intelligence agencies to disclose anything and in supporting this proposal in this way you bring new resources to their disposal.

This provides for comprehensive context instead of fragmentary stories. The benefit in terms of global unity far outweighs national security interests; there surely must be a way to keep secrets and perhaps learn something yourself, intelligence communities.

By using only the information in the public sector, no pressure is put on secrecy. This is beneficial to the intelligence community in that secrets expressed here are no longer secrets, and there is no asking you for information, nor should they be in this commission. It is fact-finding, and the report that this roundtable is mandated to produce is the consensus its members provide. Disagreements between members can be decided by a one-thousand-word statement explaining your position and why the other position is wrong, limit five to a member.

The political benefit will accrue to democratic institutions even while doing so protects their adversaries until said adversaries' children can process the data. At this point, the world will change.

Gillibrand is to run this in the Senate and Alexandria Ocasio-Cortez, AOC, in the House.

The first problem is to look at information packages like ECETI who are conservative propagandists when their mission is to seek extraterrestrial life. It may have already found them and be active in a dark-handed way in our politics.

Education is national and global, and the very act of doing this will change politics globally and for the better.

For instance, a lab has identified the gene for psychic ability. We all have the ability, but some have it more than others. With our now manipulating that gene to produce humans with a better ability to see fourth dimensionally, coupled with AI, we become intergalactic. This could happen next week; are you prepared? Do you have a plan in place?

Reduce religious nationalism and extremism. Build a scientific cadre in each country.

This will lead to a one-world government, and our own intelligence officers should be written into the program at its inception.

While there are many benefits to this narrative, without it our security state will not only be overmatched but also out of resources.

How we do this:

This idea gives you great political leverage while at the same time opens up the playing field for our own intelligence services because it becomes a force multiplier. The intelligence community is sewn into this structure at the very beginning, and while alien-to-human technology is ongoing, what lags behind is the human community's ability to deal with it. This structure proves heaven, but dynamics change with a 4D understanding. In terms of politics, there are not many religious denominations; they are all part of a common denomination. This leaves religion in place, and family traditions are honored, but there are no longer borders. This can be accomplished by funding the proposals attached.

Director of Psyops CIA,
Your technology transfer is far ahead of the cultural development needed to manage that technology.

To cross that dimensional boundary, you need a different groupthink.

Briefly explained, money and power in Earth's model are obtained through a reductionist process of war. Speaking just for our galaxy alone, there are three to six civilizations thousands to millions of years more developed than ours. They mean us no harm and not only have shown great restraint but have also made it clear they do not want nuclear weapons in space.

There are two to five times as many planets, just in our galaxy, as developed as we are. So twenty planets at our level in this model, plus the five more developed, so twenty-five accessible planets in this galaxy times 125 billion galaxies as estimated by the Hubble deep field telescope is 3.1 trillion planets. You have the technology to reach them all and none of the resources in terms of human capital, and with a reductionist model operating here on Earth, you are going to get hurt out there. All one needs to do is look at the condition of the Russian oligarchs. This is a limited view in 3D space.

What is the change in groupthink to create acceptance of a 4D model and keep control of the process and the transition? That is what these proposals are about.

The science of reincarnation is the key to opening that lock. It leaves tradition and religion without teeth. The change will occur in one generation. Getting rid of nuclear weapons is a peripheral political benefit because this requires a global constitution.

If getting rid of nuclear weapons and the pariah of religion is my hall pass to the universe, then the ones who profit from wars have just found a better game.

For extraterrestrials, you helped humans get rid of a threat. If you are transdimensional, then bounce theory says a nuclear blast does more damage in the next dimension to this one. So you also benefit.

This limited acknowledgment can be used to disarm our adversaries and then rewrite their constitutions by going in with bankers and educators. This sells because it is a win-win situation, particularly for oligarchs and industrialists and anyone who wants peace and prosperity.

Secrecy can be detrimental if it constricts development. But managing that development à la Bernays is strategic. Psychically, bandwidth increases with the number of minds, and artificial intelligence combined with the Monroe effect will produce RAIV, the remote artificial intelligence viewer, which is an essential part of emerging national security. This is a new device and a way to view and communicate with the universe.

How is a common mind not a consensus? The best form of government to facilitate that is a democracy, which allows heat and friction to dissipate more easily on social movement. A constitution is best designed not as a political document but as an operating system. Optimally reductionist political strategies would diminish, as the health of all parts reflect the health of the whole and its collective ability to defend itself.

Think of this plan as a force multiplier; however, transdimensional rules apply, so while I may not be directing you to what you want, I am trying to help you get to what you need. There is a coming transition for your group that will change nothing and changes everything simultaneously.

Interested Others

In the film acknowledgment, there is a twenty-second clip of US Air Force representatives illegally confiscating eight hundred thousand acres around Area 51, and they are being grilled by members of Congress about who authorized that action. They would be the first interested others that this proposal should be given, the ones who authorized the action to abrogate our national laws. Give this proposal to them.

Humanity needs more than technology; it needs to redefine consciousness without destroying its traditions, belief systems, and the moral compass provided within. The upload and download of a body of data are measurable and manageable by incorporating the science of reincarnation in the study of human cognition. This recognizes the common mind. This needs both your permission and support to go forward. Facilitating this development turns belief into understanding and helps stabilize a world careening out of control.

Working toward disarmament needs your monetary support.

How is what we know to be applied?

For Kirsten Gillibrand, we ask that you request the release of zero-point energy. In this ask that you will make of the intelligence agencies, ask for an encrypted control to accessible power to ensure safety. It is already on the units in the alien craft. The intelligence agencies will control the amount of accessible power per unit. Encrypted controls to accessible power ensure safety and can be achieved. Kirsten, this will disrupt the power grid economies but transform the world by eliminating carbon emissions.

From the CIA, we ask:

1. Open your files to begin to give the timeline of alien contact. This should begin to be taught in

universities. Culling the United States military for individuals with nonlocal talent should be expanded to a global program managed through and by the consortium of schools outlined in this proposal, thereby developing scope and reaching beyond your present capabilities.
2. Develop and distribute the nonlocal power sources used to power alien spacecraft. Given the fact that any power source is throttleable, it can be made safely and transform the Earth by no longer needing local resources for power. Additionally, it powers education and reduces the need to control resources that are outside current national boundaries.

For the alien managers and visitors to this planet that are extraterrestrial:

First, a proposal like this, generated organically, is evidence of your acclimation programs eliciting a response. This is an acknowledgment that we have intersections that are clearer to you than to us. It is also clear that you do not want nuclear weapons in space. We are coming to the moon, and to the South Pole, and we ask that we interact in the best possible way. It's not just the military who will run this; it will be commercial, and Earth will need to plug into the universal structure in an orderly way.

The ask, then, is that you participate with the structure outlined in these proposals with the spaceport in

Arizona connected to a consortium of schools studying consciousness.

Implicit in this request is the understanding of a higher level of space-time and the removal of any weapons that may harm ourselves and others within the terms of the larger structure of planets/federations. I do not have the right words here, but I think you get the idea.

To the alien managers and visitors to this planet who are transdimensional:

We ask you to normalize all religions using the science of reincarnation and allow us to see our past lives enough to remove the blocks that interfere with our ability to have a common consciousness that protects the whole with the individual pieces working in unison. This ability in humans in the nascent form to resonate dimensionally should be publicly acknowledged and studied.

Now there is an implicit offering of a quid pro quo in these requests because by complying, you sanction the mobilization of trillions of dollars in resources as represented by the consortium of billionaires being recruited and organized to improve the present conditions on Earth.

First, by normalizing relations in a positive way, we take the threat of nuclear war or accident off the table. Replacing it with a nonlocal power source device or zero-point energy device that powers your spacecraft, the need for nuclear weapons diminishes to a vanishing point.

Second, resources can be diverted to the reclamation of the planet. Global warming can be reversed, and all

The Applications of the Science of Reincarnation

races who use the Earth for resources can continue or improve their condition and access.

Third, commercialization of the needs of alien races creates opportunity for all.

Fourth, the technology transfer to humanity can only be accomplished with a corresponding understanding of consciousness. The science of reincarnation, as explained in these proposals, is used by the Grays in creating hybrid beings to transfer their "essence" to. It is a simple upload and download of information, and when that information has self-reference, in short, is sentient, we call that transfer to a new container reincarnation. That needs to be explained globally to advance the acclimation projects that are ongoing. This is simply a suggestion to improve the current conditions.

The gestalt of this entire effort is that by working together, we improve the conditions and result for everyone.

In fact, in explaining the 4D arc of consciousness (upload-download, life-death), you justify the entire LGBTQ community scientifically.

This is not a discussion of when life begins. This is proof life exists before conception and after death.

Scientific support for prayer and the afterlife, you must accept the whole idea or none.

Under the wokeness initiative is the most serious national security issue there is, and we do not understand the threats until we can get on a telepathic bandwidth aliens operate on.

TIFACS Policy Positions

We cannot avoid a disclosure event in the near term. But even with that, there will be no global strategy unless we implement one. Once the totality of the alien community becomes understood, we will need a state department just for off-world activities. The civilizations we see are far advanced from us and no threat unless you regard manipulating our genes for societal advancement as a threat. Once we acknowledge them, there must be a way to ask for protection and help.

There are seed projects that will grow from this funding structure that will benefit everyone everywhere. At its scientific head is Nobel Prize winner Stuart Hameroff.

At the University of Arizona, you will develop Earth's Intergalactic Information Receptor. You see, Earth needs a global government, complete with a state department, to administer what is to come. That center has to fuse thought not to a global concept but to a universal transdimensional concept.

The University of Arizona ends up at the epicenter of this plan and ultimately will be infused not just with hundreds of millions of dollars but also the best scientific talent pool the world has to offer.

Now what I am asking from Melinda and MacKenzie is not just money but to insert a new information package into the group mind. You see that information package in the consciousness proposal. When you evaluate the merits of this grant request, I would ask you to look to its fruit. What happens when politicians or clergy must address very real changes in our reality? What about issues such as abortion, apostasy, blasphemy, or clerical celibacy? The information package I am asking to insert affects every issue.

You all know that there are asks of scientists, billionaires, and educational institutions who work together in this set of proposals. We are now at the door of the CIA and our intelligence services, and there are asks of them and the alien managers who work through them at the end of this proposal. There is one other group. We are going to call them the Intelligence Energy Consortium. This is their question. Everyone who is given an ask is given a quid pro quo. So, it shall be with this group. Imagine three worlds you control the energy of. Can you do that without passing this point?

The whole can be more than the sum of its parts. To do that, the whole must protect all its parts.

This is the essence of future government policy globally.

This is scientifically proven and intrinsically understood.

You do not need the government to disclose what it knows about aliens. If you want to meet an alien, you can just go through the Gateway Process yourself. It's simply about getting on their wavelength. Would it help if I said to focus on your own theta brain frequency and accelerate it from 40 cps to 2800 cps, and then you are on the right channel? Be directed by scientists and fund research, open research. Disclosure will come from that alone.

In human consciousness, all colors are the same. Bill Nye said you enter this world in a band around the globe that dictates skin color. Near the equator is darker, yet near the poles lighter. Differentiating this is another way to keep us apart. But common consciousness seems to be what the federation at the center of the remote viewing we do is interested in.

The University of Arizona and the Monroe Institute are prime recipients of consciousness grants. After building the network in consciousness in the twenty chairs, you have a structure for greater impact in theory development and applied research.

Homer,

The strength of the pack is in the wolf; the strength of the wolf is in the pack. All of us are needed.

Chapter 16

The Common Constitution

Funding a Cohesive Program of Consciousness Study and Its Application to Societal Laws

The explanation of this proposal in a thought. Constitutions globally need to be rewritten collectively to unite a common consciousness in mankind; here is how and why.

The purpose of this proposal and its funding go to writing a global constitution that all countries can adopt one plank at a time. There are 195 countries in the world. The goal is to standardize wellness in an intergalactic environment. To be of more value in the universe, we must do better here without causing upheaval during the transition and coalescing human consciousness to a logical and caring standard. This new demountable constitution must address Ingo's points of separation and unite them.

"A larger, much larger question eventually loomed into view: why do mass-consciousness humans, as it were, mass-consciously almost "conspire" to avoid certain issues, and consistently so? My investigations into this matter have revealed

The Applications of the Science of Reincarnation

that four general areas of societal avoidance have existed for quite some time:

1. **Sexuality and eroticism.**
2. **Human psychic phenomena.**
3. **General societal love.**
4. **UFOs and extraterrestrials.**"[112]

To begin with, his points on sexuality and eroticism are LGBTQ issues, and they are all about acceptance. With the science showing we transition from life to life in different genders, then the constitution should include acceptance, support, and protection, something Denmark would easily put in as a common plank but Nigeria would resist.

As more countries accept this plank, the world unifies on a standard of general societal love. In an intergalactic environment, infighting among ourselves is counterproductive. The last dog standing is a beaten, exhausted animal ripe for picking. A pack of dogs standing together is a lot tougher to deal with. But they also protect each other.

The argument that is more important, more powerful than the arguments for white supremacy, Islam, MAGA, Christo fascism is as follows. Ready?

---

[112] Ingo Swann, *Penetration, The Question of Extraterrestrial and Human Telepathy* (Swann-Ryder Productions, 2020), Page 225

The Common Constitution

"The strength of the wolf is in the pack; the strength of the pack is in the wolf."

If you diminish any part of us, all of us are weakened. Cross support is imperative. This is not being woke; this is being societally smart.

For earth/humans, to face the universe constitutions must work like an operating system that protects itself by protecting all parts of itself. With conflict and war in place, I diminish the efficacy of the system. Protect everything and everyone, all races, religions, genders, everything. Planks that are controversial will not be imposed, but over time they will be absorbed. So TIFACS wants a standard constitution based on this science and these conditions that countries can use as a guide when writing their constitutions and align with other countries to protect each other.

General societal love can be expressed by taking the money spent on the war in the Ukraine and feed and educate the world. Bankers can affect the greatest change here by limiting resources to aggressors. The corollary here for bankers is an environment where trade goods are being opened to and from the universe, making the earth a good market to trade in, which benefits the bankers, then everyone else too.

Human psychic phenomena are common to all humans, but humans are not interconnected. Let us say there is a telepathic language aliens use that we cannot, not that we cannot have it, but we are currently kept from it. This understanding of 4D thought transcends every religion everywhere, honors and respects each religion, and acknowledges they are all in the fourth dimension, so while heaven is now proven, so, too, is the migration of the soul through different genders, different religions, and different lives. Religious laws such as apostasy, blasphemy, and celibacy all go the way of the divine right of kings and fall to an archaic past as they have no place in the future except to represent the heritage from whence we came. All those laws are now invalid scientifically.

Regarding UFOs and extraterrestrials, the acknowledgment that they are here and present, are seeking contact, and can help us. They acknowledge they could destroy us at any time by not doing it even when we attack them. Their help can heal the world.

Finally, this constitution should be written by systems engineers. The American Constitution needs to be written to remove gerrymandering. Consensus is something the common mind produces for self-protection. That imbalance that gerrymandering introduces threatens us all by keeping us apart.

Outline

This is a twenty-year development program under the College of Social and Behavioral Science to create a common constitution that is demountable. That means its planks or articles can be adopted by the 195 individual independent constitutions/countries gradually. These articles will be driven by two things: first, a best-practices model humanity creates, and then, as AI becomes more intelligent than us, we use that greater intelligence to our advantage. This allows us to continue a decentralized structure of government that harmonizes with the best way we can manage ourselves and our world.

By producing this document and publishing it, politicians globally can work in concert to produce a better local environment without having to create the best model of the constitution independently.

Incorporated into the structure of this is the beginning of intergalactic law. This common constitution project is foundational to global consensus, random number generator coherence, and remote viewing outreach. In short, this effort is the result and a manifestation of our growing global consciousness asserting itself for its own protection. This is how the Institute for Advanced Consciousness Studies (TIFACS) proposes to address this problem.

To be successful, we must conflate individual rights with individual banking rights. What is good for one is good for all. This affects third-world countries that do not have the intellectual resources to mount such a project in terms of the law, but international banking companies can impose those standards. Local production can swiftly turn to intergalactic export.

This standardization can lead to Russian and Chinese

entry into NATO but, organizationally, would have to move away from the kleptocratic models of authoritarian rule. Global consciousness is a consensus and, at its core, both democratic and self-caring. If we are to advance as a species, we must have a road map to stop feeding on ourselves. This is that road map.

At the end of the day, this change in the model of Earth's governance anticipates and incorporates intergalactic law and commerce. It is the structural change or rebirth, if you will, of the consensus of how to run this planet considering the commercial intergalactic spaceport being proposed. The model of governance we are trying to build with this proposal incorporates the consciousness of the community, so informed consensus leads the way.

The common mind operates through democratic values to produce a consensus. Our political structure must embody that scientific fact. If we do not turn a united face of humanity toward the universe, we are and will be nothing but a backwater planet with a backwater race.

This effort needs to lead with global banking standards, and those standards applying to all lead to civil rights in third-world countries. Standardize constitutions to a best-practice model and you increase productivity and become an exporter to the universe.

The simple explanation: We have made extraterrestrial contact. For us to enter an intergalactic community, we must revamp our laws to new conditions. Countries all have their own laws and comply with international treaties. The same thing will happen here.

The intergalactic laws are written by societies far in advance of ourselves. To that end, we should look to adapt to their standards and, in doing so, harmonize constitutions here to produce our best result to respond to new threats and opportunities we face collectively. A new disease anywhere infects us all.

The common mind is a consensus, which means we all agree on something, which means we have been individually polled. The common mind is democratic by collective self-defense. This principle operated between countries, and this is now morphing in a planetary way that this proposal is addressing.

To that end, we will endeavor to write a demountable constitution with an industry best-practices mindset. To try to overthrow governments here on Earth is to reduce productivity. Each article of the constitution we will produce is the aggregate of the best model we can find here on Earth and elsewhere and will be guided by artificial intelligence so as to find a way to produce the best results. Each plank can be adopted by each country as they will, with the goal of harmonizing all constitutions to produce the best result for those governed.

AI in 2035 will be more intelligent than us. It is simply an algorithmic progression. What is being proposed here is that we solicit two different algorithms for governance, as embodied by the Heritage Foundation and the

Brookings Institute, and then let AI blend the two models after it becomes smarter than us, estimated date 2035.

We are not imposing our will on the universe; the universe is imposing its will on us.

Charles Koch is being asked to help rewrite the Russian Constitution to bring it in line with Western constitutions, where leaders retire from public life rather than being overthrown or dying in office. This is not to overthrow the Russian government; rather, it is to leave the public officials in place and institute a freer model to improve commerce and health. The goal is to provide a level of benefit to their population comparable to that of Finland. While the initial ask of you is to help rewrite the Russian Constitution, the underlying ask is to help design a demountable constitution that will harmonize the world's constitutions to facilitate trade and the safety of economies.

A premise to consider: There was a TED Talk that said there were, by mathematical analysis, 227 decision-makers in the world. They act collectively and cohesively at moments, but at this moment, that is what they must do. I am asking you, Jamie Dimon and Hugh Culverhouse, to fund this. This is how collectively and cohesively we must address a situation that threatens us all—you and I and anyone else equally, regardless of station. The win-win strategy this book proposes benefits all and protects us.

Now some math: There are about 2,200 billionaires in the world. There are 227 decision-makers who determine policy—my source is a TED Talk on how that policy is made and shaped. Of the one hundred billionaires in Russia, it is estimated they have stolen $1 trillion from the Russian people. None of them understand the changing events in alien contact and emerging consciousness science. They collectivity own the media, the politicians, and the banks but have no thought to how to manage our planet and our future cohesively, given the changing conditions, threats, and opportunities. This request is to formalize a plan of action to save the world.

What follows is the TIFACS letter to Mr. Koch.

Mr. Koch,

The Institute for Advanced Consciousness Studies (TIFACS), is asking for a contribution of $20 million and billionaires' involvement in a project. I want you to help lead a rewrite of the Russian constitution. I want the Russians to know why and how, and I want their cooperation collectively and individually.

The proposal here is simple. We take all the constitutions of the world and determine which countries have the highest wellness scores—which peoples are the happiest, healthiest, and wealthiest populations—and look at common points in their constitutions and political makeup. The Scandinavian countries rank the highest; countries that have large emigration pools are typically the lowest.

I want to assure you that this is no joke, and I want to explain how, why, and who else would be involved.

This book is a proposal that includes "asks" of Melinda Gates, Bill Gates, Mark Cuban, Robert Bigelow, Jennifer Pritzker, Ron Baron, MacKenzie Scott, and Elon Musk. Collectively, they are being asked to fund consciousness science for $450 million. The money goes to colleges and universities specifically studying aspects of consciousness and participating in the symposiums run through the University of Arizona Consciousness Center, run by Stuart Hameroff, recipient of the Nobel Prize with Roger Penrose for their work in microtubules in the neurons of the brain. When I explain the how and the why, you will understand the transcendent moment we stand at.

The How: The way I want you to begin rewriting the Russian Constitution is to hire the Brookings Institute to write the preliminary draft with the instructions to make it an aggregate of the constitutions of the NATO countries. Its goal is to mimic the metrics of the Nordic data model of social welfare, which rank in the highest tier of Western countries. I want this as a starting point to present to the other Russian oligarchs how to remodel their political system.

Look to the loss of money and opportunity this last Ukrainian war cost all of you. This will continue to happen in totalitarian societies where consensus is overruled and personality is more important than the economy. There are only ninety-six to one hundred oligarchs, and presenting any plan to structure Russian politics to their economic benefit would be novel; this effort is from the consulting firm they did not ask for but need badly, not just for them but for all 2,200 billionaires in the world.

As this is being written, the number of Russian oligarchs continues to fall through attrition. This is a real application of the science of reincarnation as a new model of global governance is born. What is happening to the Russian oligarchs could happen to us all. Ben Franklin said, "We all need to hang together, or, most assuredly, we will all hang separately." That is an example of a democratic consensus.

The remaining Russian oligarchs must cooperate with this effort, or they could be the next victim of autocratic rule. The consensus answer is that the Ukrainian war should not have been fought at all. The decision-making structure needs to be changed by changing the model that produced that result. TIFACS is asking for your help and cooperation in creating a new model of the Russian Constitution. At the end of the day, this is a warning to all oligarchs and billionaires.

To be clear, this is not asking to overthrow their government. This effort should be led by the politicians themselves, but the overall structure is to foster economic stability, growth, and an orderly and regularly scheduled transfer of power.

The Why: Why they should do this is explained in this book: To fund consciousness science and the transcendent nature of what has been proven through a body of evidence, we need to expand.

1. There is a proven continuity of consciousness that transcends death, and we come back to live serial lives.
2. Aliens regularly visit Earth now, and the documentation is so overwhelming that governments classify it and mishandle it. NASA has documented over five thousand planets, and there are alleged to be between twenty-two and twenty-four humanoid species that visit Earth now. The consciousness proposal attached here addresses direct private-party communication.

This vacillation deals with their current lack of how to deal with the emerging science of consciousness.

Enterprise Risk Management: In one turn of a generation, Earth will be a net exporter of life into the universe. It will need an educated and healthy workforce to achieve that. People then become an export themselves. You may be among them.

A meeting with you and the other one hundred Russian oligarchs to instruct the politicians what to enact literally will change the world. This proposal is not to dethrone Mr. Putin. Rather, it is a suggestion for him to retire to his yacht and be left alone, with copious amounts of money. He can come back and run in an unrigged system, and once that system is unrigged, Russia can join NATO. This process begins with this proposal to you, and it will stumble fitfully to its inevitable conclusion because as much as the governments are trying to keep a lid on it, aliens are here, and our own governments are mishandling this badly. It's not Russia against NATO; it's humanity with the ability to leave planet Earth.

So, the politicians cannot make this change, the Russian army lacks the expertise, and outside pressure will not work, but those that own the system—you and the other oligarchs—can make it better for yourselves, and that will make it better for everyone else.

The lack of war should usher in a period of increased productivity for export off-planet. All the stand-alone renewables in manufacturing—eco-friendly housing and devices—are the basis for colonization. Simply look to Amazon and Tesla's space ventures, and with one technological change, we cross the universe. It will happen in my grandchild's lifetime and your next lifetime, which by all measures we can currently make will happen between five and fifteen years after your own death.

Do the one hundred of you want to come back to a robust green planet or a shit-hole planet mired in war that fosters a lack of education and stupidity? I face the same issues that I am presenting to you here personally. The ask of me is to send you this proposal, so collectively, we can be the difference.

Charles, I am asking you for your input in designing a global system of management given the enfolding events in both consciousness science and the factual basis in which we now have to acknowledge that aliens are here in our present. We are no longer 195 countries; we are now one planet facing the consciousness universe embodied by our visitors. I am asking you to do enterprise risk management on a global scale or we as a world face the consequences.

I am asking you to do that in two ways.

1. Help write a demountable constitution. Of all the people I know, you, above all, would know how to tweak and manipulate a political system. The two groups to help write the constitution based on wellness are the Heritage Foundation and the Brookings Institute, and their combined metric is what determines the best constitution.
2. To help propagate this common constitution, I want you to help facilitate the distribution of the educational material being put out by our education department. This material would go along with the constitution in giving the bottom one hundred countries of the world good, sound science for a worldview, as well as a well-designed constitution that they themselves could not produce. It also has the benefit of the interconnectivity of laws such as tariffs and trade structure.

What defines the Heritage Foundation is self-interest. The problem for the Heritage Foundation members about consciousness science is that their paradigm changes the minute they understand that their very own consciousness will transcend their own permanent death. Strangely, that brings the alien problem into a different focus because that very 4D space in which their consciousness will exist as a waveform is a dimension humanity has to cross in order to do interstellar travel. The good news is that you can make a lot of money from these developments. Here is how.

Earth, to this point, has been a zero-sum game; one wins a piece of territory, one loses a piece of territory. This opens the galaxy to exploration within seventy years, per the predictive science presented. So, you must bring order to Earth to maximize the resources, and in that, there is money to be made. Order is not enough because of the resources needed for extraplanetary growth. You need intra-planetary growth—feeding programs that have been an anathema to this foundation, like guaranteed income or universal healthcare. Note to Homer: Homer, I find humor in the fact that these policies help and protect these billionaires and their constituents. It's the right policy because it's the smart policy.

What is needed is a best-practices constitution for all countries to follow. That means opening markets here on Earth so we can access their output. A constitution that creates standard order—and this constitution should be designed on a best-practices format, just like the best practices you put into your businesses to maximize profit.

People in this model are a product. Given the training needed and resources needed in this endeavor, then wellness needs to be built into the model to increase profit. It is the resultant metric that drives this hypothesis and plan.

To increase resources and market, banking integrity for the safety of all our assets needs to be institutionalized and protected. There is much to be done. The idea is to do good while doing well.

The alternative of profiting from division and greed leaves you alone and exposed when something game-changing threatens us all. You are being asked to design a constitution that is demountable. By that, it means that any government could take an article of the common constitution you are being asked to write and put it in their constitution, in sections such as health care, education, public works, and police management.

The idea is to improve the efficiency and output of underperforming divisions in a conglomerate or underperforming countries in the world by providing them something they cannot provide themselves: the work product and analysis of the best form of self-government.

This is directly derivative of the science of reincarnation, the science of consciousness, and the science of self-care and protection. This is in the Heritage Foundation's self-interest.

The model produced by Brookings and Heritage goes back to the College of Behavioral and Social Science to be evaluated and study how it can be implemented.

By producing diametrically competing algorithms—one from Brookings and one from Heritage—we produce the bandwidth range of competing interests and then let AI act as the arbitrator to assess and explain its result.

Society will still be controlled by the one-tenth of 1 percent; however, it will be run better and more efficiently and be more profitable and more benevolent. That last one is enlightened self-interest.

TIFACS has no interest in changing politicians or who is in power. It has an interest in creating the best plan by the best minds, with all peoples contributing as to how to deal with an emergency—the warming of the planet that we, ourselves, have created by our collective actions and must defeat collectively or die together. The older people will not live to see victory. To the young ones under thirty, you are headed for seventy years in an increasingly hellish place. So the proposal of the International Association for the Science of Reincarnation (IASOR.org) here to restructure the governments of the world is nonnegotiable if humanity is to survive. IASOR.org is a consciousness science contractor.

This means that every opposition party in places where governments are repressive and unchanging should adopt this platform as a means of measured change in your country.

So, what TIFACS has done is surveyed those countries that are the best run—the ones that people want to emigrate to—and taken an aggregate of their policies, so countries where people want to emigrate from can have the same metric as those they want to emigrate to.

Look at the list of countries by global population: 116 and 117, the Congo and Norway. They are virtually the same size population-wise, but on the wellness scale, Norway far exceeds the Congo.

Look at the countries from 100–195 by wealth. Ask yourself: Do these countries have the resources to create the best-practices constitution we are suggesting? They lack the resources. That is why the help of billionaires everywhere will support this effort, because not only will this bring peace and prosperity, it will also bring job opportunities and be a market creator for greater wealth.

Constitutions impose structure. TIFACS seeks to prevent these wars as being destructive to all of us and contributing to global warming—think burning oil fields in Iraq.

Protection of the minority is the responsibility of the majority. If the generation of children were taught this in schools with a curriculum, they would not have grown up to be the adults who committed the atrocity of ethnic cleansing. There is long and accurate science that goes into the shaping of healthy ecosystems, ecosystems where dominant species have wiped out other species, leading to wasteland scenarios. That structure must be brought forth in how we govern ourselves, so the protection of minorities must be a staple in how we choose to govern ourselves; this common constitution must reflect that, and its citizens must understand the benefits of it so that they can ask their leaders to adopt it. It is those countries that show the responsibility for self-government and education that should be at the front of the line to help save us from the coming disaster.

If you are going to plan to do this, it needs to be the best plan we can make by the smartest, most vetted scientists—not corporate lackeys—that we can have.

So, freedom of speech, freedom to dissent, and protections for civil disobedience must all be incorporated into our common constitution.

These protections must be accorded to all minorities because this protects our collective diversity.

Structure for change and term limits are built into the constitutions of the "wellest" countries' constitutions.

Health care and privacy are provided and protected. A free press, an independent judiciary, and religious freedom are incorporated into the governmental and social structure. The constitution requires the government to run the education system and courses on what TIFACS is doing globally so that groups of citizens can do that work locally and are included in the education system, so we can work cohesively together to save ourselves.

This will bring the best practices of humanity to each area. The people in power will stay in power and, in the new form of government, will be just as successful as they were before. After all, the people in the "wellest" countries are even more successful than those at the bottom of the chart, so while the adoptees of our standard constitution may be afraid of the change, they simply need not be. Smart, successful people are smart, successful people.

But why should every country have to, by trial and error, find its way to the best constitution? And the guys who are corrupt have management capability. If the structure were such that instead of defense funding, it was coastal management of large projects, they would have had their wealth and more.

Defense and war are a zero-sum game; one must lose for one to win. That is why a wellness model brings more wealth to those people who understand it, operate it, and live under it. If this is the best model for each of us individually and collectively, then why isn't it the best model by metric, taught as that? Globally. Now. A place aliens may want to make their presence known because we embody more of their ethics.

I might add, common defense protocols to UAP operations would strengthen our ability to respond to threats by incorporating the standards needed to respond directly into 195 constitutions.

That is what I am asking you to do, create and give the world a document that benefits everyone everywhere. You will need the whole $20 million to be distributed to schools and organizations throughout the consortium network. Start with the twenty schools of parapsychology that TIFACS is helping to found and fund.

I hope you help.

Homer, again the devil is in the details. This is a tough one, and it is not about a common constitution at all. We are funding consciousness science. Can we as a species "fit in" with extraterrestrial life as we prove we are transdimensional? Go ahead, Homer, ask Lisa what she would do while Bart develops weapons. Our best and brightest need to explain this easily for everyone and face our reality.

So how do we structure an operating system to protect and teach us all and address the multitude of life smarter than us that calls earth home?

Start with once common document. Homer, I want Lisa's take on this. I am asking billionaires and scientists left and right for help. I want Lisa to tell me what to do.

Lisa dear,

We need cohesion on Earth to respond to aliens; if we destroy ourselves, then they do not have to.

Signatories to this common constitution agree to reduce their defense budget by 10 percent from military costs to health and welfare of the citizens costs.

Constitutional rights follow political policy which follows scientific empiricism.

The constitution is not from the UN, or a country, but TIFACS's attempt is to write a constitution that draws on the best minds to produce an operating system beneficial to all parts of the whole.

So, Lisa, could you provide planks that should be incorporated across all constitutions? You can reach us at "r/reincarnationscience."

Well, Homer, this is a transfer of money that would directly benefit you and your family.

Izvestiya Pravda. These two Russian newspapers dominated Moscow news distribution in the late twenty-first century. One was the *News* and the other was the *Truth.* The word on the street was there was no news in the truth, and there was no truth in the news. This is news to many, and it is the truth.

Chapter 17

Army Futures Command: Religion and Exopolitics

Before I leave this book, I want to let you see the universe through my eyes to help you understand what all this is about. It is overwhelming at first, and then normalcy will set in.

This chapter is for people of faith worldwide. What about God? What about heaven? What about faith?

When entering a new space, it is always best to understand the situation there. This book is about preparing humanity for that entry. Do aliens have a religion? Are all aliens hostile? What can we expect? Here is a short and quick survey that answers those questions.

As far as I can tell, there are currently eighty-two known races that either live on Earth or visit it. These races are thousands of years ahead of us in development. They interact with one another, so their belief system is much older than humans' but represented in all our religions. They communicate through a common telepathic language. This should comfort any person of faith. But why? Why are so many alien races humanoid? What do they believe?

Let us meet the eighty-two alien species, break down exopolitics, and learn how to successfully interact with the universe—all within a single chapter.

Before I begin.

This chapter is lightly footnoted. Asking me to divulge more sources is asking for my death. I want to get out of writing this book alive. Bigelow is right: the fourth dimension can be a dangerous place. It is also that way in three dimensions. The information that follows is presented using the same math we use to judge reality in other sciences: odds-against-chance calculations and then fractal patterns. Once those conditions are met, you have your truth—whatever it is.

In his book *Alien Races: All Alien Species Revealed*, Alan Fredrich mentions meeting a retired government worker who "worked in a Top-Secret organization called S.D.I. (short for Search Detect and Identify). They work very similarly to Men in Black. When there is a report about an encounter with extraterrestrial forces, they show up immediately. They ask witnesses about what they saw and if they lie or try to tell others what they just saw S.D.I. threatens witnesses or their families."[113]

Now go back to the film where this book started. The retired agency worker is describing exactly what the teacher at school experienced.

---

[113] Alan Fredrich, *Alien Races: All Alien Species Revealed* (2021), 6.

The Westall Incident 1966

On April 6, 1966, outside a school in Westall, Australia, on the outskirts of Melbourne, the country's second-largest city, a flying saucer reportedly landed near Westall High School at around 10:15 a.m. It was seen by a large number of students from both the high school and the primary school next door.

This is a widely witnessed mass event. Shane Ryan investigated the Westall Incident and interviewed one hundred witnesses, including the science teacher, Mr. Andrew Green, who went out to investigate himself and saw the saucer. He spoke clearly and publicly about what he saw. A week or two later, there was a knock on his door, and he opened it to see two uniformed Australian Air Force officers. They came around to say "We don't want you talking about what you say you saw. If you speak again to the media, we will make sure you don't teach again in this state."[114]

They had come to threaten him. There are repeated events like this.

So what I am about to tell you has been cross-checked with other accounts. Many have been abducted, and the government has interacted with multiple species. As

---

[114] *Secret Space UFOs—In the Beginning,* directed by Darcy Weir (IMDb, 2022), 1hr., 25 min., https://www.imdb.com/title/tt14998318/.

The Applications of the Science of Reincarnation

noted in chapter 3, when you do talk to an alien, they will often talk about each other.

Some are benevolent, some are evil, and others are neutral. Others are not even from our dimension. In those dimensions skin color and religion do not matter.

According to Fredrich, "But this may shock you, but there are aliens that look like us, talk like us, and could walk down the street without you noticing anything strange about them."[115]

Paul Hellyer, a former defense minister in Canada, spoke openly about alien races; therefore, among the astronauts, defense ministers, government contractors, and abductees, the following information is the most coherent way I can explain this in one chapter.

For the religious, many races believe in the Cosmic Law of One—we are all one and serve one.

> **"Cosmic Law of One:** These laws are designed to align lives to the universal frequency of harmony
>
> 1. Law of Unity: Everything is interconnected. When you are at one with yourself, you are at one with everyone. There is no separation.
> 2. Law of Vibration: Tuning the spiritual body to vibrate at a higher frequency enriches life.

---

[115] Fredrich, *Alien Races: All Alien Species Revealed*, 7.

3. Law of manifestation: Using your frequency to manifest your hopes, dreams, and desires.
4. Law of Cause and Effect: The energy/frequency you put out is the energy/frequency you get back.
5. Law of mental Clarity: Connecting to Source daily for mental clarity.
6. Law of Spiritual development: Connecting to Source daily to design your spiritual path toward Divinity.
7. Law of Physical Health: Eat right, exercise, and laugh.
8. Law of Social Interaction: Maintaining close friendships with family, friends, and making new friends does the heart good.
9. Law of Compensation: Working towards personal goals and achieving the end result no matter how challenging the path may be."[116]

As I write this book, I have found that there are more than one hundred known races, and this number will keep growing. I cannot keep up, and neither can our authorities. More resources need to be brought to this if humanity is to survive.

The following races all follow this belief system: Agarthans, Altarians, Andromedans, Twenty-Four

---

[116] Craig Campobasso, *The Extraterrestrial Species Almanac: The Ultimate Guide to Greys, Reptilians, Hybrids, and Nordics* (MUFON, 2021), 270.

Elders of Andromeda, Antarians, Apunians (from Alpha Centauri), Arcturians, Arians, Cassiopeians, Ceitans, Celestials, Clarions, Created Beings, Cyclops, Cygnus Alphans, Digital Immortals, Eridaneans, Guardians, Klermers, Koldashans, Lyrans, Alcyone Pleiadians, Pleiadians, Procyonans, Proxima Centaurians, Sagittarians, Blue Sirians, Human Sirians, Soulzars, Vegans, Venusians, Zeta Reticulans, Largan Peoples, Lion-Felines, Mantis Beings, Seventh Ray Race, Zeta Lizard Humans, E'all Reptoids, Lizardian People, Rigelians, Royal Dinoids, Saurian Beings, Jowlen, Ummites, Watchers, Moramiams, Zeta Humans, Zeta Humans-Blue and Golden Caste, Beesonites, Serpent Beings (49 races)

These races would be nonhostile and perhaps helpful.

Others compatible with the Law of One: Itipurians, the Lady of Light (transdimensional being), the Melchizedeks, Titan Sirians, Solar Light beings, Superangels, Sassani Beings (7 races)

Races that are neutral: Grey Insectoids, Sirian Insectiods (2 races)

In opposition to the Law of One: Renegade Pleiadians, Synthetics, EBEs (Extraterrestrial Biological Entities, the Grey Races, Orons, Mothman (6 races)

These races can be dangerous: Anunnaki, Dinosaurians, Draconians (Dracs), Alpha Albino Royal Dracs, Dragon Dracs, Zeta Dracs, Iguanoids, Repterrians, Reptile Imposter Humans, Small Reptoids, Bat Bowouls (11)

Total races mentioned: 75

Army Futures Command: Religion and Exopolitics

In all seriousness, how can the US government effectively deal with the entire universe? Once people of faith learn to navigate the fourth dimension, the phase lock will be broken.

Religion is an entry point to exopolitics. As I pointed out in chapter 3, anyone can talk to aliens through a telepathic language whose protocols were developed and enhanced at the Monroe Institute. Now reincarnation becomes global and societal.

I want to paint the landscape in broad strokes. Where do we fit in? What is happening? What motivates action?

According to Campobasso, "While it seems most of the universe is humanoid (and benevolent), there are also intelligent, malevolent life-forms as well, including reptiles, Greys, dinosaurian beings, insectoids, and even some humans. DNA is more precious than any alloy across the cosmos, and genetic engineering and manipulation are standard practice."[117]

Altarians are from the Aquila constellation. "Although most are light-skinned, some Alterians' skin tones range from blue, green, tan, and brown to a variety of other hues, from breeding with other human races. The skin tone spectrum is a natural occurrence within star nations."[118]

Now justify your prejudice against people of color.

---

[117] Campobasso, *The Extraterrestrial Species Almanac*, viii.

[118] Campobasso, *The Extraterrestrial Species Almanac*, 7.

The Applications of the Science of Reincarnation

Education wipes out cultural differences, and the technology transfer that may benefit us must take our cultural mores into the future. You cannot be enlightened if you are willfully ignorant.

Campobasso added, "Andromedans are blue hairless humans with three classifications, female, male, and androgynous."[119]

Here is a race thousands of years more advanced than humans. Is your prejudice to the LGBTQ spectrum going to play with this alien group, or are they going to think you are primitive, which you are? Your prejudice is also primitive considering fourth dimensional thought. Or perhaps we could be prejudiced against all aliens because they are bluish. Why not? We discriminate if they are Jewish.

There are races exactly like us to the point they could walk down the street, and you would not know it. "Altarians are indistinguishable from Earth humans, though they are taller.[120]

"Arians are indistinguishable from Asian races on earth."[121] "Ceitans are indistinguishable from earth humans…Their race resembles Spanish, Italian, and American Indian heritages on earth."[122] Campobasso

---

[119] Campobasso, *The Extraterrestrial Species Almanac*, 11.
[120] Campobasso, *The Extraterrestrial Species Almanac*, 7
[121] Campobasso, *The Extraterrestrial Species Almanac 33*
[122] Campobasso, *The Extraterrestrial Species Almanac 41*

says that Antarians and Apunians could also easily pass in the Earth's population.

Campobasso added that adult Agarthans "may choose up to five live-in relationships (bond marriages) in different households, their time divided equally. All forms of sexuality are embraced. Learning how to love unconditionally in personal and professional relationships is paramount."[123]

Are you, a human, going to tell the Agarthans, who look like humans and live here on Earth, that they are wrong about their sexuality? Is it possible that Elon Musk knows more about sexuality than a race hundreds of thousands of years older than humans?

The only way to effect a cultural change is to introduce the common man to our alien neighbors. For that to happen, CE-5 protocols and Monroe protocols must be taught globally. The book *A CE-5 Handbook: An Easy-to-Use Guide to Help You Contact Extraterrestrial Life*[124] explains this well. Go to www.ace5handbook.com or www.etcontacthub.com to get the full free PDF.

Here is a simple guide to contacting aliens: either use the CE-5 guide or listen to the Monroe tapes. Here is your introduction to aliens. Humans, these are the aliens. Aliens, please be gentle and understanding; we are novices in this endeavor. There you go—you have

---

[123] Campobasso, *The Extraterrestrial Species Almanac*, 4.

[124] Hatch, Koprowski & The Calgary CE-5 Group, Amazon, 2018, www.ace5handbook.com, www.etcontacthub.com

been formally introduced. The common man is now off and running into the universe. Anyone who does this can call themselves a psychonaut.

Courtney Brown and the Farsight Institute should be paid to contact the crop circle makers and begin a conversation about better drugs for humans. He should do this with Mark Cuban's help, whom I previously asked to do this. Governments should not have all the fun by keeping the technology and contacts a secret.

That brings me to the hardest explanation of what is happening in our sector of the universe and what is happening around us. At this point, science meets science fiction in a hard collision. So, contact with these alien groups has painted the following picture.

According to Campobasso, "The Galacterian Alignment of Space Peoples and Planets is a universal alliance of fully conscious beings that have united their planets to benefit all universal-kind."[125]

He also added, "The Star Seed Alignment is a subdivision of the Galacterian Alignment. It's one of the many established consciousness-raising programs. Universal citizens temporarily leave their lives in the stars to reincarnate on worlds in the pangs of duality, such as Earth."[126]

Let us talk about raising consciousness. We have already proven that reincarnation is real, and you believe

---

[125] Campobasso, *The Extraterrestrial Species Almanac*, 263.
[126] Campobasso, *The Extraterrestrial Species Almanac*, 263.

in heaven. It is like Santa Claus—children are taught to believe in him, but at some point, they realize it is their family getting them gifts. As a race, we have come collectively to the realization that heaven is the fourth dimension, and aliens much older than humans exist both there and here. Some wish to use us to manipulate DNA, while others at higher dimensions curate our souls—the energy that makes us who we are. As Stephan Schwartz taught me, this, too, means a shift in the metaparadigm.

For our safety, the following should be a government-sponsored program: A good strategy to save humans is for all religions to align with the Law of One. Our religions are mostly there. Then we must align our laws and actions accordingly. This means hypocrisy must go away, and the result is a better planet with better health for us all collectively.

By adopting the Law of One, our religions remain intact, and each denomination recognizes a shared foundation. This forces change in both religious and secular laws to align with these principles; otherwise, we risk hypocrisy to the very races we are trying to align with. The quid pro quo would be a defense of our planet and ourselves. To go any other way would be to sell ourselves into slavery.

This means that all religions should be proactive—have your parishioners use this system, contact dead relatives, have OBEs, and ask about new drugs or technology. You will not be able to lie; they will see your intent. So,

The Applications of the Science of Reincarnation

aligning with the Law of One means you change laws, or else you are a hypocrite and not ready to address what is before you.

So, when you die and go to heaven, they say you ascend to heaven. This higher plane/dimension is occupied by beings much older than us. One such race is the Melchizedeks. According to Campobasso, "They are indistinguishable from Earth humans."[127]

> **Belief System**: Melchizedeks is a spiritual university of the most advanced, and the Melchizedeks are known as the first order of creator sons.
> **Cosmic Agenda:** The Melchizedeks are part of the Galacterian Alignment. The University of Melchizedeks is the nucleus where programs of Living Light are dispatched to all worlds.[128]

Pause for a moment and look at the fractals of organization. The Consciousness Center at the University of Arizona is a fractal of the University of Melchizedeks. They are both studying the light of consciousness. This alignment should allow for a greater flow of information. But first, we need to be conscious of the connection and then courageous enough to design a plan to make contact within the academic community without

---

[127] Campobasso, *The Extraterrestrial Species Almanac*, 93.
[128] Campobasso, *The Extraterrestrial Species Almanac*, 93.

government help. Let the government keep its secrets and allow the information to the academic community be open and global.

Look at the fractal alignment of their stated mission and our collective religions.

That means universities globally that offers chairs in parapsychology must provide accredited courses on how to go about alien contact and monitor and categorize the responses. This will be a renaissance in human development.

This is no longer fiction; this is science.

But wait. If aliens have been here for 12,500 years or 200 million years, what about aliens who live here among us now?

They control our wars and politics, are raising us humans as a race, and are manipulating and using our DNA. We are reaching a point where we might be allowed to join the galactic community if we do not kill ourselves like the Martians. The CIA knows there was a nuclear war on Mars 200 million years ago based on the soil samples they picked up, and they do not care if you know because they let the researcher who made the discovery publish his findings.

The way to deal with this is through education—the chairs of parapsychology. It is the only way for all of us to get on the telepathic channel to save humanity, our children, and our grandchildren.

The AFC must fund the TIFACS plan. Being a global effort and accepting the new conditions, the AFC needs to lay the foundations for the war thirty years in the future. Will we be fighting each other? Will we be fighting together to survive as a race?

MJ 12 is the organization buried deep within the US Government; we discussed at the end of Chapter 6. Majestic 12 was the first 12 men appointed by Truman to oversee the alien issue. It still exists today. MJ 12 must encourage and support the TIFACS plan. The plan not only supplies added support to an overwhelming task, but acclimated humans in a gentle way to the concept of aliens and the 4th dimension. Deep in MJ 12 past the human component, that mind knows education is the only path to success. This is a path to soften the landing of the announcement of the extraterrestrial presence. The ask of you is to get TIFACS done. I do not want your secrets; I want to help Earth/humanity/the universe. Let the AFC support and help implement this plan.

Earlier, I mentioned some books that should be included in university Science of Reincarnation courses. The Science of Reincarnation is about the rebirth of the human mind and its collective societal significance. This directly leads to connecting with the universal mind through and with alien species capable of space flight as listed above in this chapter. The following books need to be included in that curriculum.

Kennedy's Last Stand: Eisenhower, UFOs, MJ-12 & JFK's Assassination by Michael E. Salla

Exopolitics: Political Implications of The Extraterrestrial Presence by Michael E. Salla
Galactic Diplomacy: Getting to Yes with ET by Michael E. Salla

TIFACS is charged with creating the curriculum explaining the new conditions and politics imposed upon us by these discoveries. Universities globally are charged with teaching this curriculum so we have a chance to save ourselves.

Chapter 18

Conclusion

Given the overwhelming evidence of extraterrestrials and transdimensionals and humans' ability to access other dimensions, there is a need for policymakers to consider the political implications. Based on this, the following five policy recommendations can be made.

"First, the quality of evidence substantiating an ET presence and clandestine government coverup has a significant degree of credibility and persuasiveness. This supports the creation of a new field of public policy, exopolitics, which would study these two perspectives in the current political climate of an officially sanctioned government policy of non-disclosure of the ET presence."[129]

This can be done by setting up a congressional committee that would be the forerunner of the cabinet position we addressed in chapter 15.

"Second, there is a need to promote official government disclosure of an ET presence and to make more

---

[129] Michael E. Salla, *Exopolitics Political Implications of the Extraterrestrial Presence* (Tempe, AZ: Dandelion Books, 2004), 47–48.

representative the policy-making process that has evolved in government responses to such a presence."[130]

While the process we outline for this eventuality, first the congressional committee, and then the cabinet post, military secrets should remain secret. The level of secrecy should be reduced, as denial is no longer a believable option and continued secrecy has become detrimental because more resources are needed to deal with this global issue.

"Third, there is a need to reveal the full nature of national security policies undertaken by clandestine government organizations in militarily responding to the ET presence."[131]

When doing this TIFACS becomes a buffer in the global acceptance of this new condition. It allows for global study and understanding instead of promoting fear.

"Fourth, there is a need to release into the public arena all knowledge about alternative energy sources that have commercial application but are withheld on national security grounds."[132]

The strength of our planet lies in our citizens.

---

[130] Michael E. Salla, *Exopolitics Political Implications of the Extraterrestrial Presence* (Tempe, AZ: Dandelion Books, 2004), 47–48.

[131] Michael E. Salla, *Exopolitics Political Implications of the Extraterrestrial Presence* (Tempe, AZ: Dandelion Books, 2004), 47–48.

[132] Michael E. Salla, *Exopolitics Political Implications of the Extraterrestrial Presence* (Tempe, AZ: Dandelion Books, 2004), 47–48.

Strengthen them and you strengthen our global defense. Withholding that information weakens all of us.

"Final policy recommendation is that there needs to be more effort in determining the extent to which congressional oversight is required for organizations created to deal with the ET presence. Evidence suggests that elected public officials including even sitting Presidents, have been denied access to information about the ET presence based on national security considerations."[133]

To achieve this, we must democratize access to alien culture through education, not a congressional battle with our intelligence agencies. This is why creating an NGO like TIFACS is imperative to our global safety.

This information will transform our culture as religion will be transformed, as will the reasons we fight each other. Ingo Swann has given us the pressure points of this transformation.

"A larger, much larger question eventually loomed into view: Why do mass-consciousness humans, as it were, mass-consciously almost 'conspire' to avoid certain issues, and consistently so? My investigations into this matter have revealed that four general areas of societal avoidance have existed for quite some time: ***Sexuality and eroticism. Human psychic phenomena.***

---

[133] Michael E. Salla, *Exopolitics Political Implications of the Extraterrestrial Presence* (Tempe, AZ: Dandelion Books, 2004), 47–48.

General societal love. UFOs and extraterrestrials. [emphasis mine]"[134]

Ingo concludes by saying "It is with good reason I believe, hitherto almost unimagined, that all four of these areas are at least linked with regard to an Extraterrestrial abductee context which is positively awash with sexuality overtones, while the psychic nature of UFO abductee experiences is visible beyond argument."[135]

Now let us make the case that humanity is being influenced by extraterrestrials that can manipulate mass opinion and make large groups of humans act against their own self-interest all the while manipulating our DNA for their benefit.

Look at the rhetoric from the 1930s of nazi Germany and compare it to the rhetoric of the 1924 US election. Code words are purifying blood, non-inclusivity, and attacking the gays and transgenders. Those policy positions are dividing us instead of uniting humanity to a common mind and a common understanding. Understanding the reality of the 4th dimension is to understand a transgender person is making a decision based on their own 4th-dimensional identity. Once that single cognitive event occurs, then heaven is not inaccessible, it becomes accessible and with it the universe and all of the alien life in it.

---

[134] Ingo Swann, *Penetration: The Question of Extraterrestrial and Human Telepathy* (Swann-Ryder Productions, 2020), 225.

[135] Ingo Swann, *Penetration: The Question of Extraterrestrial and Human Telepathy* (Swann-Ryder Productions, 2020), 225–226

The efforts of policymakers should be directed to these four points to begin to address this. These are the trigger points that allow us to be manipulated, divided, and conquered.

Aliens are here and yet the news that dominates our headlines has to do with transgender people, when the data shows we live life after life in different genders. What is more important, aliens manipulating human society or someone who still remembers and identifies with the gender of a prior life they lived? Human psychic phenomena are generally discredited, yet the TIFACS focus on supporting the Monroe Institute will in a generation address that. The opening of fourth-dimensional awareness will mute the threat of extraterrestrial presence as we a race will be able to communicate with other races on an intergalactic telepathic platform. Finally, these efforts collectively will remove the stigma of extraterrestrials and allow any one of us to communicate and interact simply by learning the protocols taught at the Monroe Institute. It will also protect us from being societally manipulated as we are now.

The most important and integral change is not the recognition of extraterrestrials but human consciousness. The best investment in our mutual defense is an investment in the research and education of human consciousness and our individual connection to fourth-dimensional activity. Every country in the world should want a chair of parapsychology located in their country because of the access it gives that country to the universe.

So how do you go about implementing a plan to address all these issues without tipping the global apple cart over? That would be the plan being advanced by TIFACS here.

An estimated total cost for the proposed program is about $450 million for the first year, dropping to perhaps $100 million each successive year. This is nowhere near $1 billion per year, and for good reason. At this stage of our knowledge, it is very unlikely that $1 billion could be spent wisely. A program audit and reassessment would be planned on a five-year basis to judge if funds spent so far were used well and would recommend adjustments up or down. A question not yet addressed in this moon-shot proposal is whether a single overarching organization would be in charge of the whole program or whether it would be distributed among several organizations. In either case, the overhead in running and tracking a $100–$450 million program is nontrivial. It would require a staff of perhaps twenty people, which would cost perhaps $5 million per year. Still, compared to an annual billion-dollar allocation, that would be in the noise.

—Dean Radin,
The Mathematics of the Science of Reincarnation

Dean's proposal is fine, but it will cost twice his $450 million because truth has consequences. So let us explain what this proposal has asked for in the simplest form.

Dean's contribution to this plan is pure research, but the need to drop to $100 million per year is wrong; the need will accelerate, as will funding requirements. Suppose, just suppose, one group of aliens declares their presence to the world and their internal language is telepathic. What is humanity's plan then, listen to the latest elected narcissistic despot and we all follow like sheep?

TIFACS has asked to fund Dean's proposal by having Robert Bigelow manage the contributions. So, to start, Robert Bigelow, because of his superior understanding of consciousness science, contributes $20 million to found TIFACS per this plan and begins to solicit $20 million from other billionaires.

There are 2,200 billionaires in the world; if he gets just 1 percent or twenty-three people, he gets $460 million, including himself, and Radin's plan is funded.

But integral to this is the Monroe portion, which leads to the involvement of the Army Futures Command whose mandate is to seek out new technology and opportunity. The commanding general of the Army Futures Command (AFC) could ask some billionaires if they would like to contribute. AFC spent $33 billion by their admission and got no harvestable new weapons systems. The results of

Radin's plan as outlined here meet the criteria that AFC is charged with finding.

The AFC spent $33 billion, $100 million would represent three-tenths of 1 percent, or 0.003 of the $33 billion. That $100 million is laughably small compared to the demonstrable harvestable benefit of the consequences of TIFACS plan. While the $450 million in Radin's plan does not expect ROI, the next $450 million reeks of it. If five billionaires come from this group, we would have our $100 million.

TIFACS is asking Gillibrand, Alexandria Ocasio-Cortez, and Ilhan Abdullahi Omar to sponsor an initiative for a new cabinet position to deal with exopolitical strategy. The afterlife is proven; read the papers from the best consciousness scientists in the world that Bigelow has assembled. Not only proven but accepted by the US Army, with protocols to navigate those extra-dimensional places. This plan expands our bandwidth.

The proposed roundtable in the previous chapter on policy lets the US government keep its secrets and allows the government to see a concise picture of our reality by assembling the information in the public sector. The information is already out there. The sponsored roundtable will organize this information.

TIFACS is at the center of raising money and creating a plan to harness human energy to create a course of study that brings resources to what our government is doing. Our government is simply overmatched and does not quite know how to proceed. Aliens are running acclimation programs; apparently, we are being helped by some alien species, but we, as a species, cannot leave money on the table in a game like this. The proposals we are advancing help everyone, including the aliens who are included in this proposal.

AOC, I want you to take the lead on this. This science is the argument to lay to rest the discussion on transgender issues, LGBTQ validation, and the end to apostasy, blasphemy, and celibacy across all the world's religions. This will take two generations to eradicate. However, the younger generation will be transformed quickly given the new understanding of data transmission. You will watch the death throes of the older ideology as that generation dies off.

Ilhan Abdullahi Omar takes the second seat. The science and the resultant policy are valid, but everyone needs to hear it from their own.

White supremacy and fundamentalism weaken us from within, so if you are going to fight the war thirty years in the future, this must be actuated now. A white fundamentalist explaining this to a Muslim community would be suspect. Omar taking the Gateway Process course and presenting it to her constituents is more successful and helps her, her constituents, and the white fundamentalists simultaneously.

You can fight all you want about differences, but we are all the same in death.

Senator Gillibrand: You can deliver billionaires to this overall project. Radin and I use different numbers because we are singing a melody in two parts. The key to all this is understanding human consciousness; I am trying to prevent the mess that this will cause.

What do the aliens get? The aliens get an Earth better than it is right now. Earth gets more resources to manage diverse genetic populations and preserve them. There are no losers here; it is a win-win.

The Veracity of This Proposal— You Can Check the Sources

This is as true as the situation was twenty years ago. Things are progressing. Let us assume the authorities did not tell the general population about this because they were scared shitless and could not do anything about the situation, and the power derived from retro-engineering alien technology was too lucrative. If deals are being made in the backrooms, a situation could occur where it would not be the first time a human has sold another human into slavery.

There can be a lot of bad endings for those who prefer fairy tales and ignorance—even the well-intentioned intelligence service needs to come in from the cold. Secrecy between branches of government can be detrimental to the entire organization.

The real test is in how we organize going forward, given how this changes everything.

Conclusion

Where are the aliens? In 3D or 4D, this question needs two answers.

In 4D, we need to fund the 4D apps using AI, which is already being done at the CIA, among others.

In 3D, the mission is probably already planned. RAIV is already happening. DNA modification using CRISPR is ongoing.

If the intelligence services/CIA do not include the collective human mind in the operation, it wastes the greatest resource humanity has.

Wouldn't you like to inform people of what science says? The standard model of consciousness will do that and change our views of death, religion, and gender.

The resultant change in global politics should be managed toward commercialization and common safety, but there are devils in the details.

Aliens will benefit as well.

We are all in this together—all of us and the aliens.

This is only about consciousness.

These need funding now.

Now let us all step back from it. At its core, this plan has made accessible to our intelligence and government leaders half of the world's wealth embodied in the 2,200 billionaires.

Once TIFACS is actuated, there is a second proposal for this. That is the International Association for the Science of Reincarnation (IASOR, IASOR.org).

This is open to everyone for free, but it accesses the

other half of the world's wealth which represents the net worth of everyone else and is accessible on Reddit at "r/reincarnationscience."

The two together represent a common consciousness. Every sect, every color, every race, and every gender is included. Space is open to everyone. How much deeper into the universe could NASA see if they could harness this power? How we are received by races thousands of years more developed than us will be based on how we treat each other here on Earth.

The goal here is symmetry. That is why the Monroe initiative is imperative. They aspire to show you an experience that proves life after death that you can manage and visit at will.

Bob Lazar said he was told that aliens regard humans as receptacles of souls. This is a correct statement but wrong information. Humans are not receptacles for souls, humans are the instruments of souls. Change the operating system of Earth and humanity becomes a galactic neighbor.

What these TIFACS proposals are about is bringing an upgrade to the present situation by bringing more resources to the effort. If this is about changing operating systems, then Earth's operating system is getting an upgrade.

Changing one part of a system can significantly impact the system's overall function because systems are made up of interconnected parts that work together to produce specific outcomes. When a part is changed, the system may need to adapt to maintain balance and continue to function. This change can affect other parts of the system through adaptability or causality. For example, in a food web, removing a producer can impact the entire web because there is less energy available.

In complex systems, change can be a dynamic, two-way process. Changes in one area can lead to changes in other areas, which can then cause further changes, and so on. This process can echo the core characteristics of complex systems, such as interdependence and unpredictability

In short, that 1 percent Monroe is after will change the entire world in a generation.

How do we manage that process? Infuse Monroe with resources and expand their reach through the Consciousness Center and the twenty chairs of parapsychology so the formalized teaching goes worldwide to all universities, and be honest about what you are doing and what it means.

All that divides us will be swept away in the tide.

The US Army Futures Command should lead this effort, being the best-run, best-placed organization to meet these challenges.

Science is the priority, safety is the necessity, and a plan like this is the expediency to protect us all, including the aliens.

We all need to do our part.

This is something billionaires must react to because you cannot control it.

To all, you now have a fully integrated plan to upgrade human culture and consciousness to a higher level of awareness, health, kindness and inclusion, and sustainability in a larger environment.

Health can be infected to harm the body, and infectious viral ideas must be laid to rest by courageous scientists before they infect the common mind. Ideas antithetical to logic must be addressed even when they contradict religious belief.

Politicians must minimize societal disruption.

Politicians must articulate and execute a thirty-year strategy, not for the next election cycle. That means taking a backseat to the scientists.

I want to step outside of the $450 million needs outlined by Dr. Radin and ask for the US government to ask Lockheed Martin, Boeing, and McDonnell Douglas to contribute here. The alien ships of the Grays are piloted by thought, and while human thought may be successful in understanding the dynamics to translate human thought into action by an intelligent machine, there needs to be an overriding structure. The connection is not just to the spacecraft but to every system in it. This goes to navigation in a dimensional manifold. This is connecting the consciousness of the pilots to the craft they will fly.

It is not just that we already have alien reproduction

vehicles (ARVs), but the questions are where we would go in them, and what would we do when we get there.

This transcendent logic-driven operation has political ramifications in LGBTQ, religious, and racial legislation. This is a political position based on science, knowledge, and truth.

How is what we know to be applied?

Kirsten, in December 2022, you were making disclosure requests in the DOD budget. What you are asking for can be improved by asking for the declassification of technology that will transform the world. On the spacecraft obtained by our government is what looks to be a transparent plastic sheet, roughly eighteen by twenty-four by one-quarter inches. We now know this is the power plant of these spacecraft, and it accesses zero-point energy. A household model would power any building worldwide for virtually no cost. In your subsequent request, this is what you should ask for. Not just the zero-point power pack but any technology that can improve the world.

In this ask that you will have of the intelligence agencies, ask for an encrypted control to accessible power to ensure safety. It is already on the units in the alien craft. The intelligence agencies will control the amount of accessible power per unit. Encrypted controls to accessible power ensure safety and can be achieved. Kirsten, this will disrupt the power grid economies but transform the world by eliminating carbon emissions.

To the hidden behemoths who control society, here

is the request for you: Allow zero-point energy to be released to the public. You can control any accessible amount of power through an encrypted design. Earth can die like so many other planets through misuse, or Earth can change for the better, and humans with it. In one movement you can wipe out the pollution of fossil fuels, end the carbon era, and create a new global industry of installing and maintaining the equipment.

There is the issue of humans out in the galaxy that may be offensive to other races, but to let Earth be destroyed when so many come here to use its resources would be galactically wasteful. A better way to manage Earth is the commercialization of extraterrestrial needs. So, while the behemoths may fear this development, it increases their wealth, power, and reach. Only when the behemoths acquiesce to this development will humanity be allowed it. So then this is a written request asking you to make this so. Universal free power.

The benefit for you of participating is to add to both your power and safety by developing resources you do not yet have and will. The game you sit at is an iteration of every game on every planet you now have access to. That is why this is a good and fair request to you, and we hope you see its merit. Because it opens other planets to you.

From the intelligence services, we ask:

1. Open your files to begin to give the timeline of alien contact; this is not asking for secrets. This

should begin to be taught in universities. Culling the United States military for individuals with nonlocal talent should be expanded to a global program managed through and by the consortium of schools outlined in this proposal, thereby developing scope and reaching beyond your present capabilities. Additionally, contact with ex-military who have died may provide useful intel.
2. Develop and distribute the nonlocal power sources used to power alien spacecraft. Given the fact that any power source is throttleable, it can be made safely and transform the Earth by no longer needing local resources for power. Additionally, it powers education and reduces the need to control resources that are outside current national boundaries.

For the alien managers and visitors to this planet that are extraterrestrial:

First, a proposal like this, generated organically, is evidence of your acclimation programs eliciting a response. This is an acknowledgment that we have intersections that are clearer to you than to us. It is also clear that you do not want nuclear weapons in space. We are coming to the moon, and to the South Pole, and we ask that we interact in the best possible way. It is not just the military who will run this; it will be commercial, and Earth will need to plug into the universal structure in an orderly way.

The ask, then, is that you participate with the structure outlined in these proposals with the spaceport in Arizona connected to a consortium of schools studying consciousness.

Implicit in this request is understanding a higher level of space-time and removing any weapons that may harm ourselves and others within the terms of the larger structure of planets/federations. I do not have the right words here, but I think you get the idea.

To the alien managers and visitors to this planet who are transdimensional:

We ask you to normalize all religions using the science of reincarnation and allow us to see our past lives enough to remove the blocks that interfere with our ability to have a common consciousness that protects the whole with the individual pieces working in unison. This ability in humans in the nascent form to resonate dimensionally should be publicly acknowledged and studied.

Now there is an implicit offering of a quid pro quo in these requests because, by complying, you sanction the mobilization of trillions of dollars in resources as represented by the consortium of billionaires being recruited and organized to improve the present conditions on Earth.

By normalizing relations positively, we take the threat of nuclear war or accident off the table. Replacing it with a nonlocal power source device or zero-point energy device that powers your spacecraft, the need for nuclear weapons diminishes to a vanishing point.

Resources can be diverted to the reclamation of the planet. Global warming can be reversed, and all races who use the Earth for resources can continue or improve their condition and access.

Third, commercialization of the needs of alien races creates opportunity for all.

Fourth, the technology transfer to humanity can only be accomplished with a corresponding understanding of consciousness and culture. The science of reincarnation, as explained in these proposals, is used by the Grays in creating hybrid beings to transfer their "essence" to. It is a simple upload and download of information, and when that information has self-reference, in short, is sentient, we call that transfer to a new container reincarnation. That needs to be explained globally to advance the acclimation projects that are ongoing. This is simply a suggestion to improve the current conditions.

The gestalt of this entire effort is that by working together, we improve the conditions and result for everyone.

So, since so many of our interests are intertwined, how should we collectively proceed for the best result for the entire group?

To all concerned, nuclear war on Earth benefits no one, so everyone should be on board to help Earth get over this dangerous hump. Earth is in development, and by all we can see, helping us get out of the periods of war would protect the resource pool for all of us. If our information is right, the Grays suffered such a calamity.

So how are we going to proceed? This question is asked not just to my human readers but my alien readers as well. If you are working to bring humanity along in development, there are moments that show greater maturity and understanding. This is one of them. Presenting a method to communicate, such as the science of reincarnation, to everyone, you emphasize the importance of global harmony as well as harmonizing humanity with the universe at the same time. You are interacting with humanity now, and the situation might be optimal for you, but cooperation would be best. This is a step toward that, a proposal coming from Earth, and just the act of creating it is the creation of a new reality.

I ask every one of you: Had you been Eisenhower, what would you have done? My own opinion is that they did the best they could in their moment, and then their moment became the structure to overcome. We cannot negotiate because we do not have the power or sometimes even the knowledge to defend ourselves. Still, by all indications, there is a structure at work that does, and remote viewing found it.

This plan is a benefit to all and a better use of resources.

I am not giving away any national secrets, but you are at a singular point. Billionaires, you are going to have to do what governments should have done. But the investment opportunity is immense; it rebuilds Earth, and it opens the universe. It changes the planet, and it changes

Conclusion

us. We cannot stop it, but we can make it better, easier, and gentler.

It is safer for alien extraterrestrials and safer for alien transdimensionals.

The aliens know that some of us humans know they are here. However, the acceptance of this fact by the public is being cultivated through acclimation programs directed by the aliens themselves. This request, then, is to direct a response the aliens would be interested in and to begin a commercialization process through the structure and process outlined in this series of proposals.

There is one corollary to this request that should be made clear. No request in these proposals is made without a corresponding benefit to the requestee. That benefit is managing your own consciousness after death and into the next life. That is right—not just reincarnation, but managed reincarnation, with the universe accessible anywhere. We started with a math proof that reincarnation in humans was real. We then found out that the Grays use a system to move their consciousness to new bodies. What we will learn in theory development is how this will lay the foundation for this being a real event. To achieve this, humanity has to move forward in a nonthreatening way to the alien consciousnesses that surround us, both in our 3D and dimensional manifold environments.

This should be something our consortium of billionaires should be actively pursuing. To that end, TIFACS is proposing an alliance between the intelligence community and our consortiums to manage that technology transfer to the world.

Within this structure we can be both harmed and helped, but we can also harm or help ourselves. This suggestion is just that, an attempt to help ourselves face the larger universe of our neighbors.

That means we should commercialize the aliens' needs. We are being used as a genetic farm by one species and threatened by another, yet protected by an over-galactic core agency protecting emerging species. We can help or hurt this web through our actions.

1. Water for mining the moon. In *Penetration,* Ingo talks about his experience watching an alien ship suck water from an arctic lake. Aliens come here and use our natural resources. Can we add value to their needs by providing services in exchange for technology? Can this be done commercially rather than through the government?
2. Genetic infusion into the Grays. There must be better ways than abducting people at night. Frankly, a more efficient way must be found for everyone, or is there something happening we do not agree to?
3. The transdimensional entities at the South Pole as expressed in the Byrd documents and other instances should be addressed if they are willing.

4. Our neighbors on the moon who do not want nukes in space. How is that negotiated? We do not own the moon. They were there first.
5. The dead Martian world inhabited by the remnants of Martian society as remote-viewed by the CIA. Also, Buzz Aldrin's acknowledgment of a monolith on the Martian moon.
6. We have a very real need to develop a political science platform at all our universities that includes exopolitics to deal with this, its intricacies, and the introduction of AI as a means for conflict resolution.

In conclusion, from the point of view of the military and special ops training, not developing the human capital in the second and third world in health, welfare, and education is leaving money on the table in the competitive game humanity has entered in an unacknowledged intergalactic reality. Not to see your weakness by being fooled is to be willingly complicit in your loss.

This begins with three empirically proven discoveries not yet processed or understood due to their transcendent ramifications.

1. We have scientifically proven a continuity of consciousness after bodily death.
2. We have scientifically proven the existence of alien life and visitors who traverse our airspace

and oceans and whose presence is current and continual.
3. There are transdimensional beings who live on or frequent the planet Earth. Anything humans do to harm the planet harms their environment.

You may disagree with all three premises, but the empirical proof is absolute. Twenty million is small enough from each billionaire, but your participation makes this a cohesive whole in terms of an intelligent collective measured response. Some may not wish to participate; these positions may be taken by other billionaires. It is not just the money you are being asked for; it is your participation/guidance on various aspects of how we as a species should respond to this new information. Bigelow should start with $20 million from each of you for a total of $44 billion.

Each of the recipient organizations, most of which are 501(c)(3)s and specific individuals within, are given "asks," producing a collective intellectual ask overall, with each having part of the whole. Individually many may come to this project reluctantly or not even know of it. But they will come because, like the ask of you, it is the right thing to do.

Given the size of the money and distribution of it in this proposal, some of you are being asked upfront to fund various projects and some of you are being asked to contact specific individuals to discuss where said contributed money should best be dispersed. Each of you donors will have specific targets within the overall plan based on your exhibited talents and interests.

In summation, you, the money, are joining the talent, meaning the scientists, to form a team to produce the most effective response humanity can have to the new conditions that are emerging, proving continuity of consciousness after death, and aliens frequenting our planet.

Your first question may well be what the afterlife has to do with aliens. Each is a fractal of nonlocal consciousness. Consciousness is what we are proposing to study. Telepathy, clairvoyance, out-of-body experiences, and the afterlife are all fractals of our connection to cosmic communication, which is a data-acquisition multiplier. The business and military applications are transcendent; they are next-generation and beyond, and some aspects of this proposal will look to commercialize them.

The Why: Why you should do this is explained in the attached proposal to fund consciousness science and the transcendent nature of what has been proven through a body of evidence we need to expand.

1. There is a proven continuity of consciousness that transcends death, and we come back to live serial lives.
2. Aliens regularly visit Earth now, and the documentation is so overwhelming that governments

classify it and mishandle it. NASA has documented over five thousand planets, and there are alleged to be between twenty-two and twenty-four humanoid species that visit Earth now. The consciousness proposal attached here addresses direct private party communication.

Enterprise Risk Management: In one turn of a generation, Earth will be a net exporter of life into the universe. It will need an educated and healthy workforce to achieve that. People then become an export themselves. You may be among them.

Now equate this to the proposal request for funding for the Noetic Institute that has its fingers in identifying the gene for psychic ability. The $450 million now is an imperative, and the government is not capable of the type of response being proposed here. If you are a billionaire, how far into your next life can your money and the science of reincarnation take you? There is an answer, and that is common intent, meaning not just your intent but the intent of others. This is explained in many religions, but they are just stories. This is about the manipulation of your own personal reality. Said plainly, reincarnation is more than just a mathematical certainty; it's funding the science to build the technology. Then let me ask you, Mr. Billionaire, where are you going?

In the proposal, we explained the need to fund that intersection between alien and human consciousness and its study.

In the proposal, we explained the need for a global cohesive response to the emerging conditions for all parties, including aliens present in our reality. This includes extraterrestrials and transdimensionals.

Here in the proposal, where we are now, this plan is a benefit to all and a better use of resources.

I am not giving away any national secrets, but you are at a singular point. Billionaires, you are going to have to do what governments should have done. But the investment opportunity is immense; it rebuilds Earth, and it opens the universe. It changes the planet, and it changes us. We cannot stop it, but we can make it better, easier, and gentler.

It is safer for alien extraterrestrials and safer for alien transdimensionals.

The simple explanation: This proposal is about monetizing and democratizing the technology obtained from the retro-engineering of crashed alien spacecraft. To that end, this is an appeal to one politician in this entire proposal series, Kirsten Gillibrand. The appeal to her is the appeal to all politicians to request the same thing of their governments.

The aliens know that some of us humans know they are here. However, the acceptance of this fact by the general public is being cultivated through acclimation programs directed by the aliens themselves. This request, then, is to direct a response the aliens would be interested in and to begin a commercialization process through the structure and process outlined in this series of proposals.

What this series of proposals represents is as follows.

Proposal 1: Produce a consensus standard model of consciousness using the same scientific protocols to produce the standard model of cosmology. The profound consequences of this act are to rewrite religion, heaven, LGBTQ understanding, psychics, and our connection to the universal mind. The implications for psychological warfare as well as for intelligence and operations are far-reaching. This would be housed at the University of Arizona Consciousness Center and led by Nobel Prize winner Stuart Hameroff. He would have to integrate the Noetic Institute model of consciousness into a unified whole.

Proposal 2: Prepare humanity for proof of aliens. Publicly analyze the intergalactic community that humanity will have to deal with. Establish a welcome center.

Proposal 3: Common constitution. Rewrite the Russian Constitution and standardize on a published best-practices constitution. Establish global educational goals with everyone having total access. This establishes the common mind, a democracy in that a common mind must come to a consensus in a 4D state. Still, some of those things are punishable legally in 3D space. What things? Racial prejudices, sexual presentation, and adherence to doctrines that have no bearing like clothing choices for women. A best-practices constitution would require this to be law.

Proposal 4: Privatization of the technology starting with a zero-point energy power source universally available.

So, regarding the structure of the legislatively required disclosure per December's budget, I say again, be guided by our intelligence community.

You do this by funding consciousness initiatives to produce an inclusive standard model of consciousness.

This is accepted given that aliens are here, there are multiple groups, they have different agendas, and rather than being a threat, they are managing our development.

This should result in the reduction of religious nationalism by introducing 4D awareness supported by AI.

More advanced societies use AI to become more spiritual, existing in some cases as not only extraterrestrials but also transdimensionals.

Using this approach, you change in one generation global attitudes at all levels and in all countries. The weapon of attack is education, global education on the science of consciousness, and through the consortium, the contributions are going to you. You get the resources of all the best consciousness scientists in the world. At the very same time, they demonstrate to off-world observers a plan for emerging growth.

Your hands are clean, sir, with a place to use other people's money to support and enhance your programs without accountability in broad daylight.

This reduces to only one reason to do this: You need more resources, which brings them.

Finally, this includes you, my reader, whoever and wherever you are: Help by doing what you can to improve your school, community, and politicians to adhere to the values TIFACS presents because you join people worldwide in protecting and providing for humanity. This is the same for all of us.

Mr. Bigelow, in all truth, I hated writing this to you. I am seventy-six; where is either one of us going? I am asking you to ask all 2,200 billionaires for $20 million each to start TIFACS. You already see who sits on the board and who does the science. Please.

It is others who must carry this forward. I want to leave a record of how to proceed with a very complex problem.

How would a society five thousand years more advanced than us handle it? That is the plan we must write and fund. So please run another contest, this one to prove aliens exist in our world and the role that they play, both extraterrestrial or transdimensional. The same prize money, the same structure.

So, we are at the conclusion. Let us sum up what this book has presented.

1. There are aliens with many differing agendas currently coexisting with us in our time-space on our planet and moon.

2. There are transdimensionals. Transdimensional technology is available to extraterrestrials.
3. Continuity of consciousness: Humans have a continuity of consciousness that transcends our own deaths. We are undergoing an evolutionary transition in our awareness of our reality.
4. Human consciousness has a common belief system that life continues after death. It is a commonality among all the Earth religions. On this point religions are all the same. Just dress and custom are different to make people be more submissive and controllable. Continuity of consciousness is a fact for us all.
5. Science is pushing governance to react to the new emerging conditions. Examples of this is the push for disclosure.
6. Our government is outmatched. Government, just who do you think is getting into those ships you have? Is the resistance to disclosure for societal control?
7. And now we have introduced Bernays's construct for societal control, which Ingo, in the last chapter, unknowingly introduced.

Let us go back to where we spoke about ECETI. Aliens are not just here but manipulating our internal dialogue by introducing factors that pit us against one another with white supremacy, discrimination, religious superiority, and fear of gender deviance, to name a few.

All impede our ability to access our common mind to only find out we are oscillating bits of energy, not color or gender or even people, but energy engaged in learning.

Systems belief, like the hard-core Trump supporters, radical Islamist militants, or born-again Christians, all are within systemic belief systems reinforced in their own echo chambers.

This is true of humanity regarding aliens; the information has not just been data-deficient, but disinformation and misinformation have obfuscated any truth.

Homer: This means people who could have helped have treated you like a mushroom. Mushrooms are kept in the dark and fed shit.

What these systems can no longer block is twofold. First, the wonderful science Bob Monroe gave us democratizes our ability to have the experience of finding out for ourselves that life continues, and alien energy and beings occupy the dimensional area we call heaven and our own skies.

Second is our own eyes as these alien creatures have less and less interest in staying unknown by their continuing and increasing exposure.

The only measured response we can have to this cultural challenge is to face it. The math shows us how.

The Monroe Institute's vision for the future is "reaching 1 percent of the world's population with direct experience of expanded awareness, so they know they are more than their physical bodies."

So, the question for my readers is a math question. What percentage of a system needs to change before the entire system changes?

I can talk to the general of the AFC, but the billionaires need to hear this. When this script flips, who is going to protect you?

If you drive by a VA facility, you will likely see a working pickup truck with an infantry badge sticker on one side and a veteran sticker on the other side. They are all 4D intelligence, these men who served.

Trained and deployed is an ethic of truth and justice that will be integral to humanity's growth and survival.

That alternative is to be shunned by more developed races who could provide help and protection if we approach this right. Very, very few get this.

The initiative must come from outside countries and outside militaries. It must be endemic. It starts with billionaires acting in concert with scientists. All egos are checked at the door.

So, Mr. Bigelow, to summarize the ask for you:

1. Donate $20 million to found TIFACS. I will donate the website, www.tifacs.org.
2. Solicit every one of the 2,200 billionaires directly to do the same. This book has the proposal for the first twenty or so of them. Simply send them this

book with a cover letter saying you are implementing the TIFACS plan as outlined and please send $20 million. The first twenty-three to respond get involvement opportunities in this groundbreaking opportunity.
3. Contact Dr. Hameroff at the University of Arizona and show him where to spend the first $50 million through his consortium, set up the panels I have laid out so the scientists themselves select projects to fund, and spend the next $400 million at their discretion per the format Radin has laid out. Connect with the Monroe Institute through this process.
4. Contact the Army Futures Command and ask for resources and help. You, sir, are in a unique position. With alien ships being run by directly connecting consciousness to the equipment then both Bigelow Aerospace and the Bigelow Institute for Consciousness Studies should provide a powerful platform for research and implementation of new technologies.

There are many billionaires I have left off this discourse. Here is what I want you to do. Today, send a check for $20 million to TIFACS c/o the Monroe Institute CEO, 365 Roberts Mountain Rd., Farber, VA 22938. Their phone number is 434-361-1500.

Then watch and wait two years. In that time read

Conclusion

Penetration by Ingo Swann, *Journeys Out of the Body* by Robert Monroe, *Mind Trek* by Joe McMoneagle, and the winner of the Bigelow Institute for Consciousness Science contest, Jeffery Mishlove's essay on "Proof of Survival of Human Consciousness Beyond Permanent Bodily Death."

So, Mr. Bigelow, you have the proposal, the website, the branding, the methods of delivering BICS, and proof of the message. Please send the CEO of the Monroe Institute a check for $20 million made out to TIFACS and decide who you want to run TIFACS. Then go out to two thousand billionaires to deal with these issues.

Again, that's TIFACS, c/o the Monroe Institute CEO, 365 Roberts Mountain Rd., Farber, VA 22938. Their phone number is 434-361-1500.

Thanks.

Homer, here is the wrap-up.

We have proof aliens have been here for at least 12,500 years and are here right now.

We see proof of aliens on Mars from pictures taken from the Jet Propulsion Laboratory.

They are not attacking us.

We are told we are being acclimated to the idea of aliens who are more developed than we are here.

We have discovered access to the fourth dimension through various means (Gateway).

We see 4D technology being used by humans.

Homer, fractal patterns repeat. One thousand years ago, humans lived in tribes. Then there were city-states, then countries, then trading blocks. The growth is outward from the seed. A unified planet, and then the cosmos, if we do not first tear ourselves apart or destroy our home world. Homer, if you feel the vibe, you are in the tribe. Humanity and earth. One.

Homer, did you call Mr. Burns for money?

About the Author

Bob Good is the International Association for the Science of Reincarnation (IASOR) executive director. This think tank promotes sound scientific information on reincarnation and the nature of consciousness. Bob previously spent forty years as an independent contractor servicing the pharmaceutical industry, where he specialized in research. Before that, he was involved with military research while on active duty. Once he began his business career, he was involved with Business Executives for National Security, a Washington-based think tank whose mission is to bring efficient business practices into the US military. Bob is passionate about providing education on the science of reincarnation and has spoken at Florida Atlantic University and the International Association for Near-Death Studies, in addition to authoring The Matrix of Consciousness Series. He has also appeared on the History Channel's show Ancient Aliens, dealing with reincarnation. IASOR's goal is to work toward a consensus among scientists on the prevailing scientific view of consciousness after permanent bodily death and reincarnation.

He writes both science and science fiction.

Visit the Website at www.iasor.org or www.TIFACS.org

Made in United States
Orlando, FL
11 July 2025